D1393869

PHYSIOLOGY OF FISH IN INTENSIVE CULTURE SYSTEMS

A service of ITP

PHYSIOLOGY OF FISH IN INTENSIVE CULTURE SYSTEMS

GARY A. WEDEMEYER

Northwest Biological Science Center
National Biological Service
U.S. Department of the Interior

CHAPMAN & HALL

I(T)P® International Thompson Publishing

New York • Albany • Bonn • Boston • Cincinnati • Detroit • London • Madrid • Melbourne
Mexico City • Pacific Grove • Paris • San Francisco • Singapore • Tokyo • Toronto • Washington

Cover design: Henry Alvarez; Alvarez Design
Illustration by: Stewart Alcorn

Printed in the United States of America

Chapman & Hall
115 Fifth Avenue
New York, NY 10003

Chapman & Hall
2-6 Boundary Row
London SE1 8HN
England

Thomas Nelson Australia
102 Dodds Street
South Melbourne, 3205
Victoria, Australia

Chapman & Hall GmbH
Postfach 100 263
D-69442 Weinheim
Germany

International Thomson Editores
Campos Eliseos 385, Piso 7
Col. Polanco
11560 Mexico D.F
Mexico

International Thomson Publishing–Japan
Hirakawacho-cho Kyowa Building, 3F
1-2-1 Hirakawacho-cho
Chiyoda-ku, 102 Tokyo
Japan

International Thomson Publishing Asia
221 Henderson Road #05-10
Henderson Building
Singapore 0315

1 2 3 4 5 6 7 8 9 10 XXX 01 00 99 98 97 96

Library of Congress Cataloging-in-Publication Data

Wedemeyer, Gary A.
Physiology of fish in intensive culture systems / Gary A.
Wedemeyer.
p. cm.
Includes index.
ISBN 0-412-07801-5 (alk. paper)
1. Fish-culture. 2. Fishes--Physiology. I. Title.
SH151.W44 1996
639.3--dc20 96-5252
CIP

British Library Cataloguing in Publication Data available

To order this or any other Chapman & Hall book, please contact **International Thomson Publishing, 7625 Empire Drive, Florence, KY 41042.** Phone: (606) 525-6600 or 1-800-842-3636. Fax: (606) 525-7778. e-mail: order@chaphall.com.

For a complete listing of Chapman & Hall titles, send your request to **Chapman & Hall, Dept. BC, 115 Fifth Avenue, New York, NY 10003.**

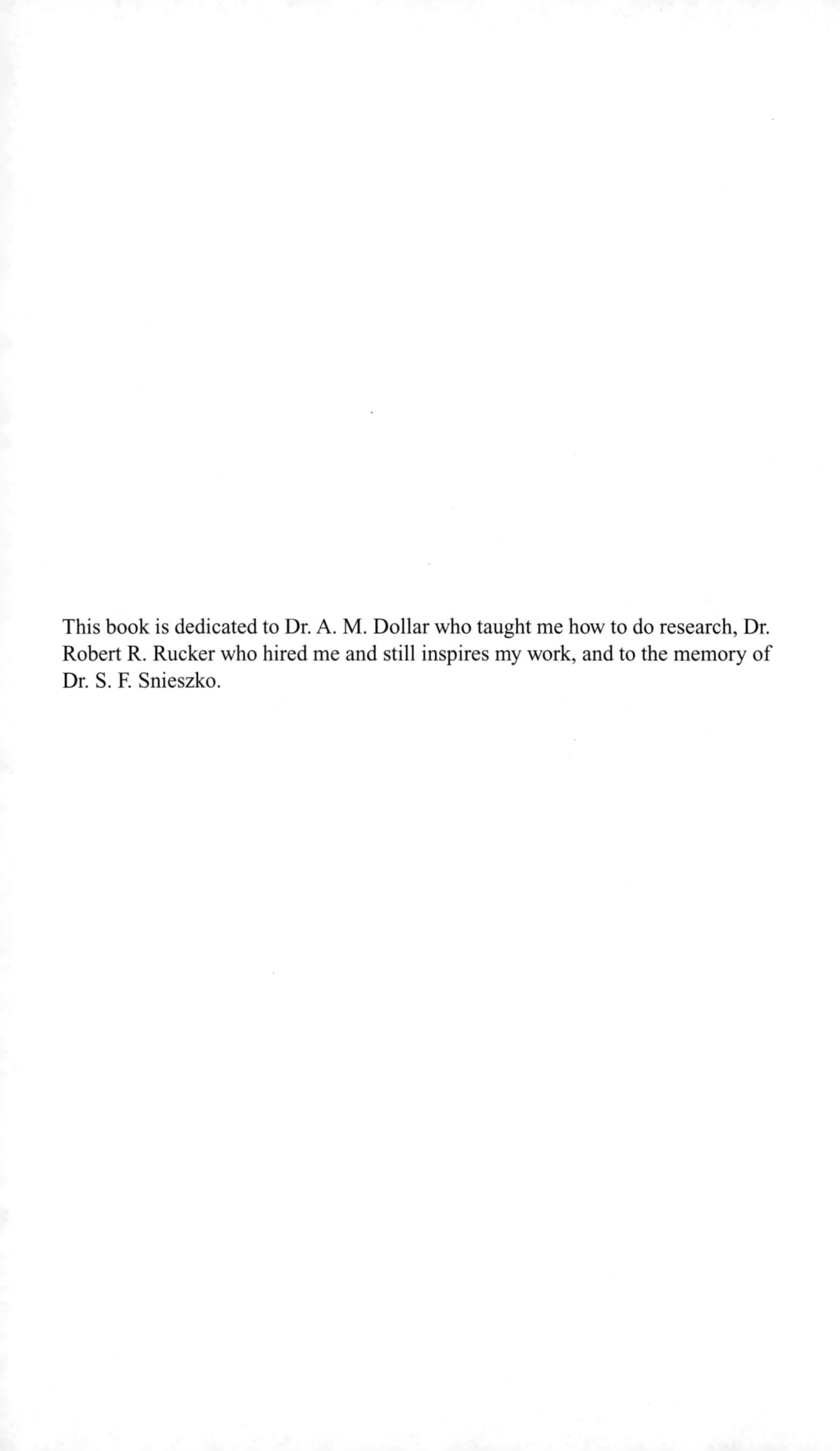

This book is dedicated to Dr. A. M. Dollar who taught me how to do research, Dr. Robert R. Rucker who hired me and still inspires my work, and to the memory of Dr. S. F. Snieszko.

Contents

Preface

Hatchery propagation for the conservation of declining fishery resources is becoming more and more intensive worldwide. Although intensive fish culture has many advantages over extensive rearing, much more careful management of the aquatic environment is required to prevent the adverse effects on health and physiological condition that would otherwise quickly occur. A more detailed understanding of fish physiology is also needed together with an improved understanding of how physiological systems interact with chemical, physical, and biological factors in the rearing environment. The effects of intensive culture on physiological condition are particularly important in hatcheries operated for the conservation of Pacific and Atlantic salmon which, in many cases, are now listed as threatened or endangered.

My main objective in writing this book was to provide practicing professionals in fishery biology, aquaculture, and natural resource management with the basic technical information needed for the rational management of environmental conditions to improve the health and physiological condition of intensively propagated juveniles. In each chapter, I have attempted to blend theory with practical applications to provide the reader with the scientific basis for managing water quality, fish cultural procedures, and biological interactions with other fish and with microorganisms to prevent the health and disease problems that can lead to costly production loses. I hope the book will also be a useful source of technical information for undergraduate and graduate students enrolled in fishery biology, zoology, and marine biology courses.

The general organization of the text is based on my previous book *Environmental Stress and Fish Diseases,* published in 1976 and coauthored with F. Meyer and L. Smith (retired). The material in the present book is based on lectures given over the succeeding two decades to practicing fishery biologists attending intensive short courses conducted by the U.S. Fish and Wildlife Service, Office of Technical Training, and to graduate students in fish pathology at the University of Washington.

Acknowledgments

First and foremost, I thank the many practicing biologists and graduate students who were subjected to my lectures on this material in the government and university fish culture and fish pathology courses I helped teach over the past two decades. Their reactions and suggestions for improvements were invaluable in helping me shape the text into its final form. Mr. Chris Horsch of the U.S. Fish and Wildlife Service Office of Technical Training gave me particularly valuable guidance in selecting and organizing the material to be included in this book. My colleagues at the Northwest Biological Science Center (NBSC)[1] provided many helpful comments on this work while it was in preparation — special thanks are due Dr. Gayle Brown in this regard. Outside reviewers included Dr. Robert Reinert, University of Georgia; Dr. Marsha Landolt, University of Washington; Mr. John Morrison, U.S. Fish and Wildlife Service; Ms. Kathy Hopper, Washington Department of Fish and Wildlife; and Mr. Robert G. Piper and Charlie E. Smith — former Directors of the U.S. Fish and Wildlife Service Hatchery Technology Development Center, Bozeman, Montana, and my long–time mentors in intensive coldwater fish culture. Mr. Harry Westers, Michigan Department of Natural Resources (ret.) kindly gave permission to share his perspectives on the historical development of intensive culture systems. Finally, I thank Drs. Alfred C. Fox and Allan Marmelstein (NBSC) for their support, encouragement, and intellectual leadership while this book was in preparation. Our many discussions of basic biology over the years gave me valuable insights into the role of hatchery propagation in the conservation of the world's fishery resources.

1. Formerly the National Fishery Research Center, US Fish and Wildlife Service.

List of Scientific Names

Throughout this book, fishes discussed are cited only by common name. The scientific names, taken from the *American Fisheries Society Common and Scientific Names of Fish Species from the United States and Canada,* 5th edition (1990) are as follows:

American shad	*Alosa sapidissima*
Atlantic salmon	*Salmo salar*
Brook trout	*Savelinus fontinalis*
Brown trout	*Salmo trutta*
Channel catfish	*Ictalurus punctatus*
Chinook salmon	*Oncorhynchus tshawytscha*
Chum salmon	*Oncorhynchus keta*
Coho salmon	*Oncorhynchus kisutch*
Carp	*Cyprinus carpio*
Cutthroat trout	*Oncorhynchus clarki*
Cod	*Gadus macrocephalus*
Goldfish	*Carassius auratus*
Grass carp	*Ctenopharyngodon idella*
Herring	*Clupea harengus harengus*
Lake trout	*Salvelinus namaycush*
Largemouth bass	*Micropterus salmonides*
Menhaden	*Brevoortia tyrannus*
Milkfish	*Chanos chanos*
Northern pike	*Esox lucius*
Pacific salmon	*Oncorhynchus spp.*

Pink salmon	*Oncorhynchus gorbuscha*
Rainbow trout	*Oncorhynchus mykiss*
Sea bream	*Sparus aurata*
Sea bass	*Dicentrarchus labrax*
Sockeye salmon	*Oncorhynchus nerka*
Steelhead trout	*Oncorhynchus mykiss*
Striped bass	*Morone saxatilis*
Striped marlin	*Makaira audax*
Striped mullet	*Mugil cepharius*
Threadfin shad	*Dorosoma petenense*
Tilapia	*Tilapia spp.*
Walleye	*Stizostedion vitreum*
White catfish	*Ictalurus catus*
Yellowtail	*Seriola spp.*

PHYSIOLOGY OF FISH IN INTENSIVE CULTURE SYSTEMS

1 Introduction

HISTORICAL PERSPECTIVE

Modern–day fish culture traces its origins to ancient China and the Roman Empire where food fish were apparently first produced in static or low flow ponds. Under such so–called extensive rearing conditions, fish are grown in an environment similar to their natural habitat with no outside food or aeration. In extensive rearing, the water is required to perform several functions: provide physical living space for the fish, supply dissolved oxygen (DO) from the atmosphere, dilute toxic metabolic wastes, and serve as the medium in which food organisms for the fish are naturally propagated. The necessity for the water to perform all these functions understandably limits the biomass of fish that can be produced. In the extensive pond culture of carp, for example, several hundred kg per hectare is a typical upper limit (Table 1.1). If the fish density is increased in an attempt to increase production, the availability of naturally propagated food organisms will usually become the first limiting factor. This limitation can be overcome by pond fertilization or supplemental feeding of artificial diets and production can then be increased about ten fold to several thousand kg per hectare (Table 1.1). If the stocking density of the fish is raised still higher, their rate of oxygen consumption will begin to exceed the pond's ability to provide DO. Supplemental aeration can be used to overcome this limitation, but then the capacity of the water and associated aquatic plants to dilute and assimilate toxic metabolic wastes will become a limiting factor. One final fish density increase will still be possible, however, if a flow of water is provided through the pond. At the point that fish production begins to depend on flowing water, it usually begins to be thought of as intensive

TABLE 1.1. Example fish densities (biomass) that can be achieved in the pond culture of carp under various gradations of extensive to intensive culture (data compiled from Westers 1984).

Culture Method	Fish Density		Biomass Produced (kg/HA)
	(kg/m³)	lbs/ft³	
Pond (*extensive conditions*)	0.04	0.002	200
Fertilized pond	0.2	0.01	1000
Fertilized pond, supplemental feed	1.0	0.06	5000
Supplemental feed, water-flow through pond (*intensive conditions*)	400	25	2,000,000

rather than as extensive culture. Under intensive culture conditions, annual fish production in the same pond can be increased to a million or more kg per hectare (Table 1.1).

Intensive fish culture has many advantages over extensive rearing. For example, the water volume is now required to provide only physical living space for the fish. Its flow through the ponds, raceways, or tanks (often termed *rearing units*) is used to deliver the required amount of dissolved oxygen. Metabolic wastes are simply flushed away. Artificial diets formulated to meet specific nutritional requirements and fed under controlled conditions provide the food supply. At a given feeding rate, the fish density (and therefore annual production) that can be achieved becomes limited mostly by the rate of water flow through the rearing units rather than by the water's volume and surface area. Intensive culture generally also requires less space than extensive culture methods and a greater degree of control over rearing conditions is usually possible. Tanks or raceways sheltered from the weather can be constructed and the water supply can be heated, cooled, filtered, treated with ultraviolet (UV) light to inactivate pathogens, or circulated through biofilter systems to remove ammonia and then reused. Feeding can easily be mechanized and automated. For these reasons, the trend in fish culture worldwide has been toward more intensive conditions, particularly for salmonid aquaculture in developed countries, but also to some extent for warmwater species such as carp and catfish (Tucker and Robinson 1990).

Although intensive culture greatly increases the potential fish production, capital and operating costs are also greater and economic success depends on achieving growth and survival rates that are high enough to offset the increased expenses. Intensive culture methods require far more disease control to prevent costly production losses than is practiced in extensive systems. Careful management of the rearing environment is necessary to prevent the adverse effects on fish health and physiological condition that would otherwise quickly occur because of the more crowded conditions. In the case of anadromous salmonids produced by conservation hatcheries, these adverse effects can be insidious because they are often not manifested until after seawater entry. Successful intensive culture thus requires a more detailed understanding of the physiology of the fish to be produced than is the case with extensive rearing methods. A good starting place for developing this understanding is to consider the physiological demands made by normal conditions in the aquatic environment and relate these to the additional chemical, physical, and biological challenges imposed by rearing conditions in intensive culture systems.

NORMAL CONDITIONS IN THE AQUATIC ENVIRONMENT

Because fishes are aquatic and poikilothermic, their life processes are carried out under physicochemical conditions that are in many ways totally outside the life experience of a homeothermic fish biologist evolutionarily adapted to conditions in the atmosphere. For example, the aquatic "atmosphere" (water) delivers both food and oxygen to fish and dilutes their toxic metabolic waste products as well. As shown in Table 1.2, the underwater environment is also physically more restrictive and chemically more variable than the terrestrial environment experienced by fish biologists. In very large bodies of water such as the oceans, fish may also experience relatively uniform chemical conditions but in estuarine areas, and in smaller lakes and ponds, wide variability is more the norm. Freshwater fishes in particular are routinely subjected to variations and extremes in the concentrations of dissolved oxygen and other chemical constituents that for practical purposes are totally outside the experience of terrestrial animals. Oxygen depletions are not unusual and at times respiration can be difficult. Life under water also subjects fish to challenges from physical factors such as supersaturated gases and water pressure changes that are usually not even a consideration in the life experience of a fish culturist.

In addition to the necessity to cope with challenges that result from chemical variability, the physiological systems of fish must function under somewhat restrictive physical conditions. Fish conduct respiration in contact with an "atmosphere" of gases dissolved in the water that surrounds them. However, the concentrations of the oxygen and other gases dissolved in water is very different from the air utilized by the fish culturist. Water is basically an oxygen–poor environment compared to air. The solubility of oxygen in water is relatively low and even in cold, freshwater, the DO rarely exceeds 10–12 mg/L. In warm water and in sea-

TABLE 1.2. Comparison of conditions in the aquatic and terrestrial environments. Note that in general, life processes under water are carried out under more restrictive and variable conditions.

Environmental Condition	Aquatic	Terrestrial
Oxygen	Low and variable, 0–12 mg/L	Nearly constant, 300 mg/L
Pressure	Variable	Nearly constant
Temperature	Variable	Variable
Chemistry	Variable	Constant
Density	High; 800× air significant energy cost to "breathe"	Low; no significant energy cost to breathe
Viscosity	High; significant energy cost to swim	Low; no significant limitation to movement

Source: After Wedemeyer et al. (1976).

water the oxygen available to fish is still lower. Overall, the DO available to fish is <5% of the atmospheric oxygen available to meet the metabolic needs of the fish culturist. In addition, the density of water is relatively high (about 800 times greater than air), and a significant number of calories from the diet must also be expended simply pumping enough water over the gills to extract the amount of oxygen needed to support life. In contrast, the oxygen concentration of the atmosphere is about 300 mg/L, oxygen depletions are not a practical consideration, and the caloric energy required to breathe the air to extract its oxygen is minimal.

An additional physiological problem results from the water flowing over the gills to support respiration. The limited DO requires fish to have relatively large gill surface areas in order to expose enough blood to the water to achieve adequate gas exchange. Active teleosts such as salmonids can have gill area ratios as high as 10 cm^2/g body weight (Smith 1982). The water flowing over the gills for gas exchange thus unavoidably also dialyzes other substances dissolved in the blood. Although this greatly facilitates the removal of metabolic wastes, it also means that osmotic forces are continuously acting to change the chemical composition of the blood and tissues. In freshwater fish, water continuously diffuses through the gills diluting the blood. To compensate, freshwater fish produce large amounts of urine—an average of three times that of humans on a ml/kg body weight basis. However, blood electrolytes are also unavoidably excreted in the process. These ions must be continuously replaced, either from the diet or from the water, by active transport through the gills. In marine fish, the osmotic gradients are reversed and water steadily diffuses out through the gills because of the higher concentration of salts dissolved in seawater. To prevent dehydration, marine fish must continuously drink seawater and physiologically distill it. In either case, the energy needed to regulate the chemical composition of the blood and tissues (osmoregulation) can consume an additional 3–5% of the daily caloric intake.

Another challenging feature of life in the aquatic environment is the high energy cost of swimming. The viscosity of water is about 800 times greater than air and fishes must expend as much as several percent of the total calories available from the diet simply to overcome frictional drag. Since drag increases exponentially with velocity, higher swimming speeds become extremely energy demanding.

Temperature changes are one of the few environmental conditions experienced by fish that are generally less severe in the aquatic environment than in the terrestrial world. However, when water temperatures do become unfavorable, it is often not possible for fish to obtain shelter from the heat or cold in the manner of land animals. In addition, fish are temperature conformers (poikilothermic) and important physiological processes such as antibody production, feeding, digestion, and growth can be adversely affected by unfavorable water temperatures.

In summary, fish carry out their life processes under variable and restrictive physicochemical conditions that in many ways are totally outside the life experience of the fish culturist whose physiological systems are adapted to the stable and less restrictive conditions of the atmosphere. Although fish are evolutionarily adapted to these environmental conditions, this does not imply the absence of energy drains. An understanding of the challenges faced by fish under normal conditions in the aquatic environment is the basis for developing priorities and limits for the additional physiological challenges imposed by conditions in intensive culture systems.

THE INTENSIVE CULTURE ENVIRONMENT

Under the rearing conditions required in intensive aquaculture, challenges from crowding, water chemistry alterations, fish cultural procedures such as handling and disease treatments, and biological interactions with other fish and microorganisms are superimposed on the normal physiological demands made by the aquatic environment itself (Figure 1.1). These additional challenges can severely tax the ability of physiological systems to compensate. Like other animals, fishes can usually survive unfavorable conditions for limited periods by expending energy. As illustrated in Figure 1.2, however, the adverse effects on fish health caused by stressful environmental alterations increase very rapidly when acclimation tolerance limits are approached or exceeded. The debilitating effects of such stress can lead to mortalities that in turn are likely to increase exponentially with only arithmetic changes in the environmental alterations in question, for example,

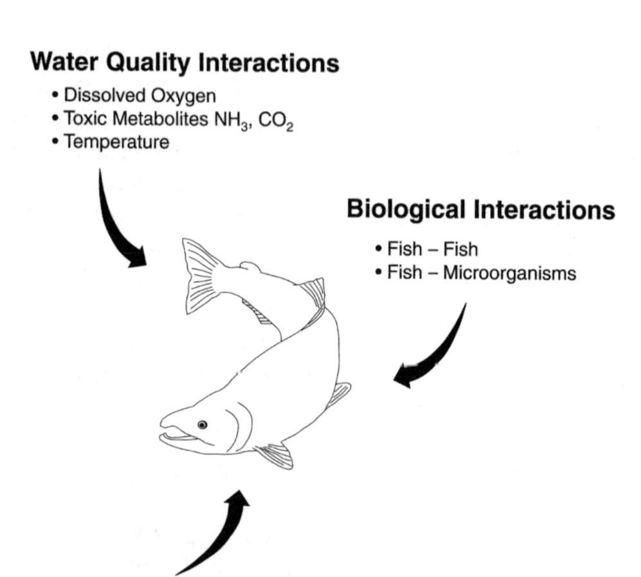

Figure 1.1. Interactions with chemical, biological, and physical factors in the rearing environment that are particularly important to the health and physiological condition of fish in intensive culture.

temperature by degrees or dissolved oxygen by parts per million. Acute or chronic stress factors such as handling or crowding that exceed physiological tolerance limits will be manifested relatively quickly as reduced survival. In the case of sublethal stress, the energy drains may still be manifested as adverse effects on physiological condition.

One of the more important consequences of poor physiological condition is impaired resistance to infectious diseases. All types of fish cultural enterprises—conservation hatcheries or commercial aquaculture—must constantly guard against the costly production losses caused by fish disease problems. The difficulty of the control measures required depends on the intensity of the culture methods being used. Intensive culture systems typically operate at high fish densities and may recirculate at least part of the water—both factors that increase the difficulty of disease control. Since these systems also have higher operating costs, even small increases in mortality result in relatively higher expenditures per fish produced.

Of all the rearing conditions affecting disease resistance and other aspects of health and physiological condition, water quality conditions, fish cultural procedures, and biological interactions between fish and aquatic microorganisms are

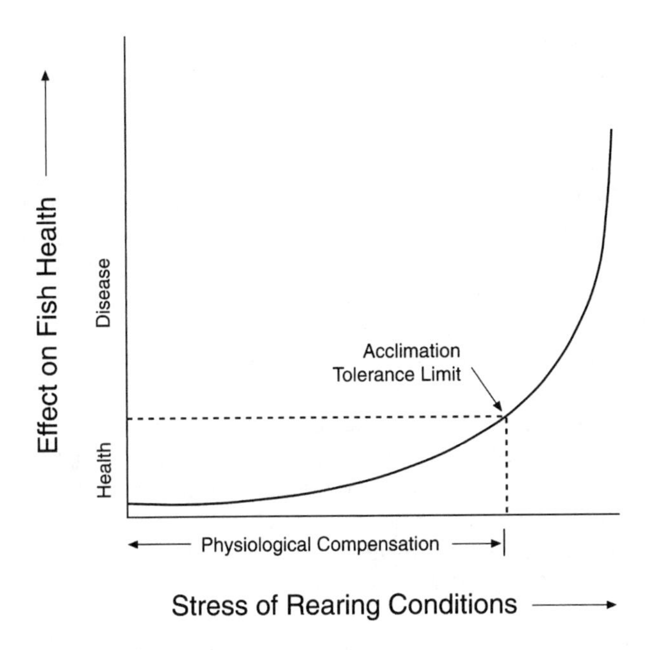

Figure 1.2. Diagrammatic interpretation of the effects of environmental stress on the health and physiological condition of fish in intensive culture. When fish are stressed beyond their acclimation tolerance limits, health is debilitated and problems with disease occur (redrawn from Depledge 1989).

probably the most important. Fortunately, many of these environmental factors are under the control of the fish culturist—at least to some degree. Managing the fish–environment–pathogen relationship to provide low stress rearing conditions should be given the highest priority in the operation of any aquacultural facility.

Water quality is usually the first place to start in efforts to provide a fish–environment relationship that will promote health and physiological condition. However, the physiological tolerance of fish to water chemistry conditions is affected by a number of variables and it is not a simple matter to identify specific constituents that will provide the optimum environment under all circumstances. For example, fish can often tolerate changes in water quality that occur separately (e.g., oxygen depletions or temperature increases) that would quickly become lethal if they occurred simultaneously (e.g., oxygen depletions and temperature increases). In the case of anadromous salmonids, seemingly minor alterations in water quality that cause no adverse effects on growth or survival during hatchery rearing can result in normal appearing smolts with inhibited migratory behavior and decreased seawater tolerance. Sublethal contaminant exposure is particularly insidious in this regard.

Although fish can often survive adverse water quality conditions because of their ability to physiologically compensate, enlightened fish health management does not use these capabilities as an excuse for allowing adverse water quality to develop or persist. Instead, knowledge of acclimation tolerance limits should be used instead to establish priorities and set limits for the rearing conditions that will promote and optimize health and condition. A detailed discussion of water quality effects is given in Chapter 3. Of necessity, the recommendations are given as individual limits of physiological tolerance but will provide the overall chemical and physical conditions needed to minimize stress and promote health and resistance to infectious diseases.

Biological interactions between fish and other organisms in the rearing environment must also be effectively managed if infectious and noninfectious fish disease problems are to be minimized. Of the many biological interactions that potentially can occur, the most important are the behavioral conflicts that result from interspecific competition and dominance hierarchies within the rearing unit itself and the interactions of fish with aquatic microorganisms. Although the physiological effects of behavioral conflicts on health and physiological condition are important, most of the effort in fish health research pertains to the relationships between fish and facultative or obligate microbial pathogens as affected by water quality and the stress of hatchery practices. Understanding the fish–pathogen–environment relationship is very important in fish culture and is the basis for rational management. Epizootics of infectious diseases are not always single–caused events; that is simply the result of pathogen exposure. Although epizootic diseases can result from the simple introduction of virulent pathogens, experience has shown that the interrelationship of other factors such as diet, water chemistry alterations, and the stress of hatchery procedures are usually also involved. This concept was first pictured diagrammatically by Snieszko (1974) as a set of circles representing the fish, the pathogens or parasites, and the aquatic environment. If the three circles intersect, conditions are favorable for an outbreak of disease. The host, pathogen, environment relationship can also be illustrated using a semi-quantitative equation:

$$H + P + S^2 = D$$

where H is the host, P is the pathogen, S is the stress caused by environmental factors, and D is the resulting disease.

Stress is expressed as an exponential quantity to account for the fact that its debilitating effects on fish health are more than additive, especially as the acclimation tolerance limits of the fish are approached or exceeded. To prevent disease problems, the additional demands on physiological systems imposed by the stress of intensive rearing conditions must be carefully managed. In hatcheries using water reuse systems, it may be practical to control both facultative and obligate pathogens by treatment methods such as ozonation or UV light disinfection. In the more common flow–through situation, the pathogen load in the rearing environment still may be indirectly under the control of the fish culturist. Practical biological and chemical treatment methods for the control of fish pathogens currently thought to be associated with adverse environmental conditions are discussed in Chapters 5 and 6.

Fish cultural procedures are the third component of intensive rearing systems that can cause stress severe enough to compromise the immune system and reduce resistance to infectious diseases. An understanding of the physiological effects of crowding and disease treatments is particularly important in this regard. In turn, this knowledge can be used to manage the fish–pathogen environment relationship to provide the low stress rearing environment required to prevent costly production loses. Of all the hatchery practices involved in fish disease problems, crowding—that is, approaching or exceeding the maximum space density limit (fish weight/unit water volume)—is probably the most commonly encountered in intensive rearing systems because of the relentless pressure of economic considerations. In the case of anadromous salmonids, crowding can have particularly insidious consequences: normal growth and development during freshwater rearing but reduced survival in the ocean. Although increased raceway loading densities result in the release of proportionally higher numbers of smolts, their marine survival is somewhat reduced. At present, it appears that low or intermediate raceway loading densities may yield an adult return as good as hatcheries operated at their maximum capacity. These and other hatchery practices having particularly adverse consequences for anadromous salmonids are discussed in Chapter 4.

REFERENCES

Depledge, M. 1989. The rational basis for detection of the early effects of marine pollutants using physiological indicators. Ambio 18:301–302.

Smith, L. S. 1982. Introduction to Fish Physiology. TFH Publications, Neptune, New Jersey.

Snieszko, S. F. 1974. The effects of environmental stress on outbreaks of infectious diseases of fishes. Journal of Fish Biology 6:197–208.

Tucker, C. S., and E. H. Robinson. 1990. Channel Catfish Farming Handbook. Van Nostrand Reinhold, New York.

Wedemeyer, G. A., F. P. Meyer, and L. Smith. 1976. Environmental Stress and Fish Diseases. TFH Publications, Neptune, New Jersey.

Westers, H. 1984. Principles of Intensive Fish Culture (A Manual for Michigan's State Fish Hatcheries). Michigan Department of Natural Resources, Lansing, Michigan.

2

Basic Physiological Functions

INTRODUCTION

The effects of rearing conditions on the health and physiological condition of fish in intensive culture are best understood holistically—as an aspect of the biology of the whole animal (*sensu lato*). Although the anatomical and physiological systems of fishes are similar in many ways to those of all animals, the adaptations that allow fish to live underwater make some of their life processes unique and quite different from those of animals living in the terrestrial environment. Of these, the parr–smolt transformation, and the processes of respiration, osmoregulation, and feeding, digestion, and excretion are particularly important to understand in managing rearing conditions to meet physiological requirements. The stress response and its effects on the immune protection system have especially important implications for disease prevention through management. Knowledge of these processes is also important to the rational design of life-support systems that will meet physiological requirements in fish transport operations. This chapter presents basic information on the functioning of these physiological and anatomical systems under normal conditions. This information will place in context the material in later chapters on managing the interactions of fish in intensive culture with the chemical, physical, and biological components of the rearing environment that can cause pathophysiological problems.

RESPIRATION AND OXYGEN CONSUMPTION

The term *respiration* refers to the exchange of gases (mostly oxygen and carbon dioxide) between the cells of an organism and its environment. In unicellular animals, the required amount of gas exchange can be accomplished by passive diffusion alone. In complex organisms such as fish, both a specialized anatomical structure (gill) for gas exchange, and a gas transport system (blood and circulation) are required to supply sufficient amounts of oxygen (O_2) to the tissues and remove carbon dioxide (CO_2). The fish gill also performs other functions, which will be discussed later, such as the exchange of ions between water and the tissues needed for osmoregulation and acid/base regulation. An understanding of the processes by which the respiratory system obtains, transports, and exchanges oxygen and carbon dioxide between the water, blood, and tissues is particularly relevant to the rational design and operation of intensive culture systems that will meet physiological requirements and promote consistently high levels of health and condition. All aspects of respiration are important but in the crowded conditions typical of intensive culture systems, effects on gas exchange can lead to immediate mortalities.

The amount of oxygen that must be transferred across the gill from the water and supplied to the tissues is substantial. For active coldwater fish such as salmonids, even the resting oxygen consumption rate can be 100 mg O_2/kg body weight/h or more. In actively swimming fish, the respiratory system must provide oxygen at rates as high as 800 mg O_2/kg/h (about 20 ml O_2/min) and correspondingly large amounts of carbon dioxide must be removed. As discussed in Chapter 1, however, water is basically an oxygen–poor environment where the maximum dissolved oxygen rarely exceeds 10–12 mg/L. In seawater, the high dissolved salt concentration reduces the available DO to a maximum of only 8–9 mg/L. Thus fish must move relatively large volumes of water over their gills to extract enough oxygen to support life. For resting salmonids, gill irrigation rates on the order of 5–20 L H_2O/kg of body weight/h are fairly typical (Heisler 1993). At rest, most fish can move the required amount of water over their gills by pumping with mouth and opercular movements. The mouth and opercules operate as force and suction pumps that work out of phase so that a relatively steady flow of water is produced. For fish under hatchery conditions, ventilation rates in the range of 40–60/minute are fairly typical producing a water flow over the gills of 5–20 L/kg/h as mentioned previously. Because of the high density and viscosity of water, the energy cost of gill ventilation is at least 10% of the oxygen consumed. Active fish such as salmonids, sharks, and tunas can achieve the required water flow over the gills by ram ventilation—simply opening the mouth when swimming. For example, Pacific salmon employ ram ventilation when swimming at speeds higher than about 1 body length/second. Some sharks are limited to ram ventilation and must swim continuously to live. In either gill ventilation method, it is theoretically possible for up to 80% of the DO to be utilized because gill anatomy is arranged to achieve countercurrent flow—the flow of water over the gills is opposite to the flow of blood through the gills. The actual oxygen utilization is species dependent, averaging 30–40% in trout, 70% in tuna, and 70–80% in carp (Randall and Daxboeck 1984). By comparison, the human lung removes only about 25% of the oxygen in atmospheric air.

As water passes over the gills, its dissolved oxygen diffuses passively through the thin (10–20 μm) epithelial cells of the secondary gill lamellae and into the blood. Once in the blood, most of the oxygen diffuses into the erythrocytes where it is bound by hemoglobin (Hb) and circulated to the tissues by the cardiovascular system. A unique feature of the physical chemistry of the Hb molecule is that increased acidity $[H^+]$ decreases its binding affinity for O_2 molecules (the Bohr effect) and in some species, its maximal capacity to bind oxygen (the Root effect). Thus, as blood circulates through the capillary beds of the tissues, the acidity caused by the carbon dioxide being produced weakens the Hb–O_2 bond thereby facilitating the release of O_2 molecules, which then passively diffuse into the cells where the oxygen tension (P_{O_2}) is low. At the same time, carbon dioxide diffuses out of the tissues into the blood—driven by its own concentration gradient. Unlike oxygen, most of the CO_2 simply dissolves in the plasma and is circulated back to the gills in its bicarbonate form (HCO_3^-). As the blood passes through the gills, the enzyme carbonic anhydrase rapidly dehydrates the HCO_3^- ions to molecular CO_2, which then diffuses out into the water. Because the residence time of a unit volume of blood within the gills is only a few seconds, this enzymatic reaction has to be extremely fast. The diffusion of CO_2 out through the gills is also a rapid process because of the high Pco_2 gradient between the blood and the water. Thus, while the blood oxygen tension may vary by 100 mm Hg or more, the blood CO_2 concentration both remains low and changes very little. The Bohr effect, particularly in active coldwater fish, tends to be large (begins at low blood CO_2 levels). In aquaculture systems for example, the Bohr effect will begin to impair the oxygen transport of salmonids if the dissolved CO_2 concentration in the water is allowed to rise to only 20 mg/L. The commonly cultured warmwater fishes (tilapia, carp, channel catfish) are generally less sensitive to the dissolved CO_2 concentration but it is still good practice to manage rearing conditions to prevent CO_2 from accumulating in the pond water.

Together with the effect of carbon dioxide, lactic acid production can cause blood acidity increases and impair blood oxygen transport. The most common cause is excessive swimming activity (due to excitement and stress), which has resulted in an accumulation of lactic acid in the blood and tissues because of an oxygen debt in the white muscle. As a hypothetical example, only a few percent of the total hemoglobin would be saturated with oxygen if the blood pH declined to 6.0 from its normal range of 7.8–7.6.

The normal function of Root effect hemoglobins is to serve as molecular pumps delivering oxygen to the eye via the choroid rete and, in physoclistic species, filling the swim bladder via the rete mirabile. The latter function is unimportant in salmonids, which are physostomes and fill the swim bladder by swallowing air. However, the normal oxygen tension in the salmonid eye also exceeds that of both the blood and the water, suggesting that Root effect hemoglobins do play a role in these fishes. Sublethal exposure to heavy metals such as cadmium or mercury is known to adversely affect the normal functions of Root effect hemoglobins but the significance of this for the health of fish in intensive culture is unknown (Condo et al. 1985).

For fish in intensive culture, problems with reduced oxygen transport to the tissues due to the Bohr or Root effects are usually caused by either the acidosis resulting from high blood lactic acid concentrations (hyperlacticemia) or high blood CO_2 concentrations (hypercapnia). Common causes are excessive swimming activity due to excitement or low DO conditions that have been allowed to develop. In addition, the use of pure oxygen aeration to achieve higher loading densities (during rearing or transportation) can result in supersaturated DO levels with hypercapnia as a side effect because the high DO suppresses the gill ventilation rate. This results in blood CO_2 accumulation and increased arterial P_{CO_2} tensions. Significantly, blood oxygen transport may be unaffected because the higher arterial P_{O_2} offsets reductions caused by the Bohr effect. In addition, hypercapnia will compromise oxygen transport to the tissues only if the resulting acidity overwhelms the normal blood buffering capacity and respiratory acidosis occurs. In fish culture situations with good water quality conditions, problems with reduced oxygen transport because of the Bohr effect probably stem from metabolic acidosis due to the lactic acid produced as a result of excessive swimming activity (Figure 2.1). The magnitude and relationship of Bohr effect respiratory stress to actual CO_2 and DO concentrations in the water were first documented by Basu (1959), in terms of the DO required to provide sufficient oxygen to the tissues to support a moderate swimming activity level (Figure 2.2). As shown, this minimum amount increases from about 6 mg/L, if little or no CO_2 is present, to more than 11 mg/L if the dissolved CO_2 concentration rises to 30 mg/L. Thus, the usual recommendation that (salmonid) fish will have adequate oxygen as long as the DO does not fall below about 80% of saturation must be qualified. If dissolved CO_2 levels are not kept well below 30–40 mg/L, the blood oxygen carrying capacity will be depressed to the point that even high DO concentrations may be insufficient to prevent tissue hypoxia.

Respiratory stress from the Bohr and Root effects can be minimized by careful fish handling to reduce excitement and swimming activity, and by a water exchange rate or aeration system design that will rapidly remove dissolved CO_2 as well as provide an adequate supply of dissolved oxygen. In practice, these are two

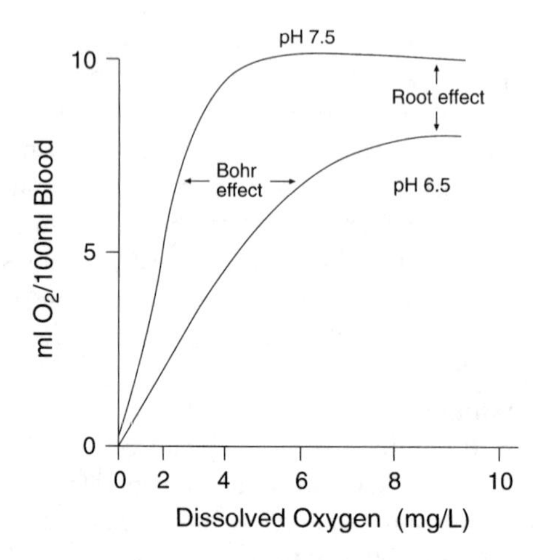

Figure 2.1. Effect of decreased blood pH due to hypercapnia or hyperlacticemia on the O_2 carrying capacity of salmonid hemoglobin.

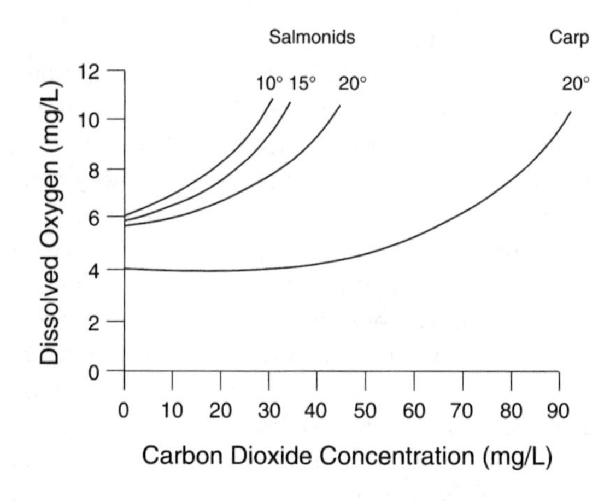

Figure 2.2. Dissolved oxygen concentrations needed to support a moderate degree of swimming activity as the amount of dissolved carbon dioxide increases. The curve for goldfish illustrates one reason for the much greater tolerance to CO_2 of warmwater fishes (after Basu 1959).

of the most important considerations in meeting the physiological needs of fish in intensive culture.

The rate at which fish consume the oxygen dissolved in a hatchery water supply is also an important consideration in managing intensive culture systems. Oxygen consumption determines such basic parameters as the water exchange rate required to support the desired loading densities and the amount of aeration needed in fish transport operations. As mentioned, salmonids under raceway conditions can consume as little as 100 mg O_2/kg/h to as much as 800 mg/kg/h or more depending on their swimming activity level, water temperature, time since last feeding, and degree of excitement or stress. Hormonal mechanisms are used to control the rate of oxygen consumption and match it to the metabolic demands resulting from exercise, stress, or the water temperature. The respiration rate of either cold– or warmwater fish is not stimulated by blood CO_2 increases as it is in terrestrial vertebrates, but rather by declines in the DO concentration. For example, when fish are stressed by handling, adrenaline and other catecholamine hormones that increase both the amount of gill perfusion and the capacity of erythrocytic hemoglobin to transport oxygen are produced. As a side effect of the branchial vasodilation, the normal osmotic influx of water increases dramatically and then must be excreted. The resulting diuresis can be dramatic and some of the blood electrolytes are unavoidably lost in the copious urine produced. If the diuresis is prolonged, life–threatening ionoregulatory disturbances can result. The delayed mortality that sometimes occurs a day or two after fish are handled and transported (often termed *hauling loss*) is largely caused by this phenomenon.

The oxygen consumption of fish in intensive culture systems can be increased both by fish cultural procedures and by natural processes. Of these, stress resulting from handling, increased swimming activity resulting from excitement, and the natural processes of feeding and digestion are probably the most important. For example, when juvenile steelhead trout are stressed by handling, their oxygen consumption can immediately more than double and remain high for an hour or more (Barton and Schreck 1987). Increased oxygen consumption caused by excitement and stress is also responsible for the rapid drop in DO that commonly occurs immediately after fish are loaded into transport tanks. If O_2 aeration is available, the hauling tank water should be supersaturated to a DO of 14–16 mg/L to help compensate for this initial oxygen demand. If only compressed air is available, starting the aeration system 5 to 10 minutes before fish are loaded to ensure that the water is saturated will help to some extent.

The natural processes of feeding and digestion also dramatically increase the O_2 consumption of fish because the caloric costs of digestion, absorption, and assimilation can amount to as much as 40% of the resting metabolic rate (Brown and Cameron 1991). The extent of this effect on oxygen consumption, termed the *specific dynamic action* of food (SDA), is not always fully appreciated because feeding is such a routine operation. For salmonids, channel catfish, and tilapia, it is prudent to plan on an oxygen consumption increase of 40–50%, or more, for several hours each time the fish are fed. Data showing the effect of feeding on the oxygen consumption of channel catfish is given in Figure 2.3. A practical consequence of the SDA is that fish should not be handled or transported immediately after feeding because the added excitement and stress may increase their O_2 consumption to the point that the aeration system cannot maintain an adequate DO. Conversely, withholding food for 24–48 hours prior to handling or transportation will prevent this effect and dramatically reduce O_2 consumption rates.

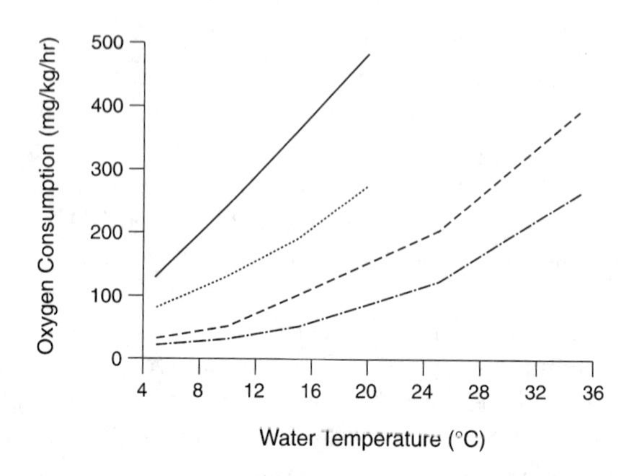

Figure 2.3. Effect of feeding on oxygen consumption. Rates shown are typical of channel catfish before (– · – ·–) and after (– – – –) feeding. Rates for fasted trout (——) and salmon (· · · ·) included for comparison.

The other major factors in intensive culture systems that can be managed to affect oxygen consumption are water temperature and swimming activity. Warmer water temperatures increase O_2 consumption rates by simply increasing the overall metabolic rate. In the case of swimming activity however, oxygen consumption increases only if the muscular exertion consumes sufficient blood O_2 to decrease Hb saturation. In rainbow trout at rest, only 60% of the gill lamellae are perfused with blood (Randall and Daxboeck 1984). The muscular exertion of rapid swimming stimulates the release of adrenaline and other catecholamine hormones into the circulation. Together with the increased gill perfusion that occurs, adrenergic stimulation of the red blood cell Na^+/H^+ exchange process takes place, which increases the intracellular pH. The Bohr effect is diminished and blood oxygenation and delivery of O_2 to the tissues are both enhanced (Thomas and Perry 1992).

The actual magnitude of the effects of temperature and swimming activity on oxygen consumption was first documented by Brett (1973) for Pacific salmon held under controlled conditions. The baseline curve in Figure 2.4 illustrates the effect of temperature alone. As expected, warmer water does increase O_2 consumption to some extent. The effect of swimming, however, is more dramatic. Burst swimming is particularly energy intensive because frictional drag is so high. The swimming activity levels of fish in intensive culture systems are usually much lower. Oxygen consumption rates that can be expected for juvenile salmon and trout reared under raceway conditions are tabulated in Table 2.1 for the range of water temperatures of interest in coldwater fish culture. Similar data for the channel catfish are given in Table 2.2.

In salmonid culture, the water exchange rate in the raceways is usually adjusted such that the oxygen consumption rate of the fish does not reduce the DO at the outflow end below about 6.0 mg/L. Aeration systems can also be used to increase

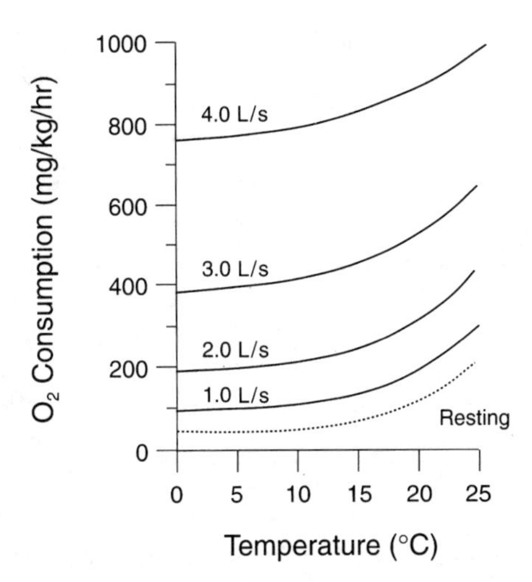

Figure 2.4. Effects of swimming speed (body lengths per second) and water temperature on the oxygen consumption of sockeye salmon. Rapid swimming dramatically increases oxygen requirements. Lower curve shows increase due to water temperature alone (after Brett 1973).

TABLE 2.1. Oxygen consumption (mg/kg/h) of juvenile coho and chinook salmon under average hatchery raceway conditions.

Fish Size (grams)	Oxygen Consumption at Water Temperature of (°C)			
	5	10	15	20
0.1	222.8	420.4	594.3	803.0
0.5	163.0	307.6	434.9	587.7
1.0	142.5	268.9	380.2	513.7
1.5	131.7	248.6	351.4	474.8
2.0	124.6	235.1	332.4	449.1
3.0	115.2	217.3	307.2	415.1
4.0	108.9	205.5	291.5	392.6
5.0	104.3	196.8	278.2	375.9
10.0	91.2	172.1	243.2	328.6
25.0	76.3	144.0	203.6	275.1
50.0	66.7	125.9	178.0	240.5
75.0	61.7	116.4	164.5	222.3
100.0	58.3	110.1	155.6	210.2

Source: Calculated from $O_2 = KT^nW^m$ (Liao 1971).

TABLE 2.2. Typical oxygen consumption rates of channel catfish (mg/kg/h) under pond culture conditions as a function of water temperature and fish size (data compiled from Tucker and Robinson 1990).

Fish Size (grams)	Oxygen Consumption at Water Temperature of (°C)				
	2	10	18	27	35
23	23	42	76	148	268
45	22	40	73	141	166
114	19	35	64	115	224
227	16	29	52	102	184
454	11	22	39	76	139

carrying capacity and in some cases liquid oxygen is employed to increase the DO to 14–16 mg/L (hyperoxic conditions). In fish transport systems, oxygen consumption rates are usually high and variable due to excitement and stress, and pure oxygen is usually provided to maintain the DO near saturation. Physiological criteria for transport aeration systems are further discussed in Chapter 4.

If the DO is not replaced as rapidly as it is being consumed by the fish, an oxygen depletion will occur. Unlike terrestrial animals, the breathing rate of fish is stimulated by falling DO concentrations, not by rising CO_2. Species such as trout, carp, and catfish respond physiologically to declining DO levels by first increasing the gill ventilation rate, using the mouth and opercules, and the blood flow through the gills by increasing the cardiac output and blood pressure. In salmonids, even moderate DO depletions result in dramatic increases in the gill ventilation rate. These actions initially increase oxygen uptake but the faster water flow also reduces the proportion of the DO that can be extracted during each pass over the gills. As the DO declines, the amount of oxygen transferred into the blood during each pass will also decline—from its theoretical 80% maximum to as little as 15% (Smith 1982). Also, the energy cost of moving more and more water over the gills increases dramatically—from the resting rate of about 10% of the oxygen extracted to as high as 70%. As a result, the energy expended to obtain oxygen steadily increases as the amount of oxygen dissolved in the water decreases and the arterial blood oxygen tension declines. When the arterial blood oxygen decreases to the point that Hb in the erythrocytes (red blood cells) is less than about 60% saturated, adrenaline and other catecholamine hormones are released that dilate the branchial vasculature and stimulate Na^+/H^+ exchange by the erythrocyte membrane, increasing the intracellular pH. Through a complex series of events, changes in Hb–O_2 affinity and capacity mediated by the Bohr and Root effects enhances both oxygen transfer at the gills and oxygen unloading at the tissues (Thomas and Perry 1992).

If the dissolved oxygen falls below about 5 mg/L, salmonids become anorexic—presumably a behavioral response to prevent the normal increase in oxygen consumption that would occur due to feeding and digestion. In salmonids, the bioenergetic cost of obtaining and utilizing oxygen begins to exceed the energy that can be derived from its consumption at a DO of about 2 mg/L, and unconsciousness, and eventually death will result. Many of the warmwater fish important to aquaculture can continue to survive for several hours even if the DO drops below 1 mg/L, but eventually the tissue hypoxia that occurs will result in unconsciousness and death in these species also.

In aquacultural practice, it is often desirable to temporarily decrease the rate at which oxygen is consumed by the fish. Fortunately, many of the same biological and environmental factors that act to increase oxygen consumption can also be manipulated to lower it. Reducing the water temperature (hypothermia), fasting, and using anesthetics to reduce swimming activity, excitement, and stress during hauling are the most common. These fish cultural procedures will be discussed in more detail in Chapter 4.

BLOOD AND CIRCULATION

The blood and circulatory system transport not only the respiratory gases, but also such vital constituents as nutrients, metabolic excretory products, and the antibodies involved in disease resistance. A number of problems with the health and physiological condition of fish in intensive culture are directly or indirectly due to pathophysiological changes in the blood and circulatory system. For example, gill parasite infestations can result in compressed capillaries and other damage to the anatomical structures of the secondary lamellae (Rousset and Raibaut 1984). The resulting dysfunctional respiration can be the basis for the overall effects of these parasites on health and physiological condition. Environmental gill disease is another good example. Knowledge of the blood and circulatory system can be helpful in understanding the effects of several of the infectious and non–infectious fish health problems faced by fish in intensive culture systems.

The fact that fish have a "two–chambered" heart and a single flow of blood through the gills to the trunk muscles, viscera, and return has important implications for fish culture. As one example, this anatomical arrangement limits the role of blood pressure in the physiological compensation for changes in environmental conditions. Although the teleost heart functionally has four chambers, there is only one ventricle where the blood pressure can be increased. The lack of the second ventricle present in higher animals means that after blood leaves the heart, its pressure and oxygen content steadily decreases as it flows through the vascular system into the capillary beds of the trunk muscles and visceral organs. In most fishes, a third capillary bed exists: the hepatic or renal portal system. In either anatomical situation, however, the blood pressure steadily decreases as it flows out of respiratory system (gills) and into the visceral organs and trunk muscles. Thus, organs and muscles progressively further from the heart are supplied with blood containing progressively less oxygen and progressively higher concentrations of metabolic waste products. Figure 2.5 is a simplified block diagram that illustrates the major features of the fish circulatory system.

The trunk musculature consists of both red and white muscle. The capillary systems supplying the white skeletal muscles are surprisingly sparse but the usually smaller amount of red muscle is well perfused. Red muscle is rich in myoglobin and produces energy aerobically from both lipids and glycogen. It is used for swimming activities, such as positioning during foraging, and for cruising. White muscle is also used for continuous swimming and for the fast accelerations required for prey capture or predator avoidance. White muscle produces energy from glycogen, which it can metabolize anaerobically into lactic acid, if necessary, thus temporarily producing energy at a faster rate than oxygen can be supplied. Repayment of this so–called oxygen debt (oxidation of the accumulated lactic acid) can require 24 hours or longer in salmonids.

The maximum blood pressure (BP) occurs during contraction of the ventricle (systole). In salmonids, the systolic pressure reaches about 125 mm Hg and it falls to nearly zero during relaxation (diastole). The small residual pressure is maintained by the elasticity of the cardiac muscle connective tissue. After the blood reaches the ventral aorta, the elasticity of the arterial wall helps to maintain pressure and the diastolic BP can be as high as about 90 mm Hg. As blood circulates through the gills, however, the very high frictional losses in the capillaries of the

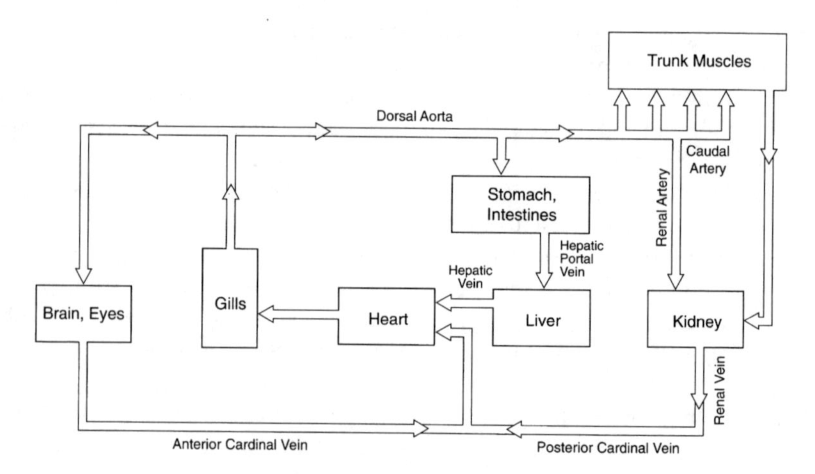

Figure 2.5. Simplified block diagram showing the major features of the fish circulatory system. Organs and tissues further from the heart must function with progressively less oxygen. Note the separate circulation of freshly oxygenated blood from the gills to the brain and return.

secondary lamellae decrease the BP as much as 40–50%. Most of the remaining pressure is expended achieving an adequate flow of blood through the arterioles and capillaries of the trunk muscles, visceral organs, and through separate circuits to the brain. Venous return from the trunk muscles, stomach, and intestines is directed through the kidneys and liver on its way back from the heart (hepatic portal and renal portal veins). After the blood has passed through these capillary beds, its pressure has been reduced to only a few mm Hg or is subambient. A pulse pressure (systolic minus diastolic) may be absent. At this point, blood flow may be assisted by the contractions of the trunk muscles during swimming. The remaining BP is usually just sufficient to achieve flow into the sinus venosus where the blood is collected and pumped into the ventricle by atrial contractions. The time required for the blood to completely circulate varies with size and species but is typically only a few minutes in salmonids.

Blood flow and pressure are regulated to meet changing physiological requirements by altering either the volume of blood pumped by each contraction of the heart (stroke volume), the rate at which the heart beats, or the resistance to flow through the vascular system. The details of how vascular circuits are altered to redirect blood flow are not completely understood but hormonal and neural control systems are undoubtedly involved. For example, adrenaline, acetylcholine, and vasoactive peptides affect blood flow and pressure by contracting or dilating the vasculature. Because resistance to flow through blood vessels is proportional to the fourth power of their radius, even small changes in diameter can have large effects on blood pressure.

The cardiovascular system of fishes lacks sufficient capacity to simultaneously supply the blood flow to the skeletal muscles required for active swimming while also providing the substantial flow of blood to the viscera required for digestion, assimilation, metabolic maintenance functions, and, for fish in seawater, osmoregulation. In resting chinook salmon for example, the normal gastrointestinal blood flow consumes 30–40% of the cardiac output and visceral flow nearly doubles immediately after feeding (Thorarensen et al. 1993). If the fish then begins to swim actively, hormonal mechanisms constrict vessels supplying the viscera diverting blood flow to the skeletal muscles to supply their oxygen and nutrient needs. If active swimming is prolonged, nutrient absorption from the gut (caloric energy intake) will decrease and growth will eventually be reduced.

In some species, adrenergic stimulation of the cardiovascular system (by adrenaline and other catecholamine hormones) causes the heart rate to increase to support swimming activity. In salmonids, however, the primary increase is in the stroke volume, although the heart rate does increase to some extent. The resting cardiac output in a 1 kg chinook salmon with an average hematocrit (30%) is about 30 ml of blood/min (Thorarensen et al. 1993).

Physical conditioning by forced swimming against increased raceway water velocities has long been considered as a method to increase the ocean survival of hatchery produced anadromous salmonids. For example, Atlantic salmon smolts subjected to forced exercise for several weeks prerelease had a greater adult return than unexercised smolts (Wendt and Saunders 1973). However, the higher rearing costs due to increased water use and the reduced growth and food conversion resulting from the forced exercise regimen have usually made any gains problematic. This may have been due to the fishes' inability to maintain the gastrointestinal blood flow required for normal growth at the forced swimming speeds that were used. It is now known that unexercised salmonids can swim continuously at only 1–2 body lengths per second without reducing intestinal blood flow and therefore nutrient absorption and growth (Davison 1989). Fortunately, exercise training eventually results in increased hematocrit and blood oxygen carrying capacity, which in turn, facilitates oxygen transport to the tissues and lowers overall oxygen consumption. As fish become physically conditioned, they may be able to swim against raceway water velocities higher than 1–2 body lengths per second without affecting growth and metabolic maintenance functions because of less need to redirect blood flow away from the viscera to perfuse the musculature (Thorarensen et al. 1993). Thus, young–of–the–year striped bass exercise–conditioned for 60 days at swimming speeds up to 2.4 body lengths per second showed significantly improved muscle development suggesting that survival in the wild might also be improved (Young and Cech 1993).

The chemical composition of fish blood is similar to that of other vertebrates in that it is about 80% water, 18% proteins, and 2% other solutes. The other solutes consist of about half inorganic salts such as Na^+ and Cl^- and half small molecular weight organic molecules such as glucose. Fish must regulate the water and most of the solutes in the blood within fairly narrow concentration limits in order to achieve the degree of homeostasis required to maintain life. For example, the blood buffering system maintains the slightly alkaline pH range (in salmonids) of about 7.4–7.6. The implication of blood chemistry regulation for fish culture is that significant departures from normal can be diagnostic of problems with health and physiological condition. Fish blood chemistry information is also helpful in the assessment of contaminant effects and for disease pathogenesis studies. Defining normal blood chemistry values is more difficult than with higher animals and tests made on clinically healthy fish held under known conditions of diet, temperature, and water chemistry must usually be used for comparison. Detailed clinical chemistry methods and normal range estimates for fishes important to aquaculture can be found in Wedemeyer et al. (1990) and Stoskopf (1993).

Chemically, blood is a fluid medium consisting of cells suspended in a complex solution of water and dissolved substances termed *plasma*. The cells constitute up to one-half of the total blood volume and are commonly classified into two groups, red cells (erythrocytes) and white cells (leukocytes). Both cell types are formed in the hematopoietic tissue of the kidney. The volume of blood in fish is quite low, averaging about 3% of body weight compared to the 5–6% that is normal in mammals.

Erythrocytes, which serve to transport oxygen, make up the majority of the cells. Fish red cells are nucleated, which confers a life span of about 1.5 years as compared to about 110 days for (mammalian) nonnucleated cells. Typical red cell counts average 1 to 2 million per mm^3 of blood in salmonids (Bowser 1993).

Leukocytes are far fewer in number than red blood cells, averaging 10,000–15,000 per mm^3 of blood in salmonids. They are differentiated in the kidney into several specialized types, usually classified as lymphocytes, neutrophils, and thrombocytes, which perform a variety of vital functions. Some of the circulating white blood cells are usually in an intermediate stage of development, but a sequence for their formation has not been widely accepted. A typical differential white cell count in salmonids would be lymphocytes, 90–98%; neutrophils, 1–9%; and thrombocytes, 1–6% (Wedemeyer and Yasutake 1977). In channel catfish at least, there are two subpopulations of lymphocytes: one group associated with cellular immunity (T cells) and one with the production of humoral antibodies (B cells). Neutrophils, so–named because they stain to the same degree in both acid and basic dyes, are the chief phagocytic leucocytes. They are polymorphonuclear and play a role in the inflammatory response. Thrombocytes function primarily in blood clotting, the mechanism of which resembles that in mammals where the cellular element is the platelet. Thrombocytes are nucleated and typically smaller than lymphocytes but may not be observed in blood smears taken from fish which have not been stressed or injured. Blood clotting time in an undisturbed salmonid is normally several minutes but blood from a stressed fish will clot in a minute or less (Wedemeyer et al. 1990). As expected, the thrombocyte count increases correspondingly as clotting time decreases.

Another group of white cells, the tissue macrophages, also exists. These cells also have phagocytic ability but are not usually found in the circulation. Tissue macrophage functions include chemotaxis, phagocytosis of microbial pathogens, and pinocytosis. Pinocytosis, the mechanism by which extracellular fluids are engulfed by cells, also provides a mechanism for the uptake and degradation of droplets of toxic organic chemicals such as petroleum hydrocarbons and pesticides.

Although not all the functions of fish leukocytes are understood, their importance in dealing with infectious agents is widely accepted. For example, following invasion of fish tissue by bacterial pathogens, some degree of inflammatory response usually occurs, together with an attempt to isolate the affected area. Tissue macrophages and circulating neutrophils, chemically attracted by the inflammatory response, begin to ingest (phagocytize) the invading pathogens. One of the mechanisms by which stress impairs disease resistance is to suppress this aspect of phagocytosis (Barton and Iwama 1991). After phagocytosis, the macrophages attempt to enzymatically destroy the engulfed bacteria using lytic enzymes, H_2O_2, and free radicals (discussed in Chapter 5). If the intracellular killing is unsuccessful, the bacteria will continue to metabolize and reproduce and the phagocytic cell may be killed instead.

If the fish survives long enough, antibodies (specific agglutinating proteins) against the infectious agent will be produced and appear in the lymphocytes, blood, and in the mucus layer of the skin. Unfortunately, fishes tend to produce antibodies relatively slowly, particularly at colder water temperatures. In the case of many infections, antibodies will not be produced in time and drug and chemical therapeutants will be needed to control epizootics. If the fish have been previously immunized, antibodies may already be available or may be produced rapidly enough to provide protection.

OSMOREGULATION

With the exception of marine elasmobranchs (sharks) and the hag fish, which have body fluids with almost the same osmotic concentration as their environment, the composition of fish blood and tissue is not usually the same as the dissolved materials in either freshwater or seawater. In addition, the dissolved substances in the blood are separated from the aquatic environment by only the very thin gill epithelium. The gill is permeable to allow the diffusion of oxygen, but water molecules are also small enough to easily pass through, although at a slower rate than oxygen. In the absence of gill lesions, such as from bacterial infections or parasites, the only dissolved materials really excluded from diffusing through the gill membranes are high molecular weight molecules such as proteins. This means that an inward or outward concentration gradient is always present for some of the substances in the blood and tissues.

The concentration gradient between substances dissolved in the blood and separated from the aquatic environment by a semipermeable membrane results in a continuous influx of water by osmosis in freshwater fish, and a continuous efflux of water from fish in seawater (Figure 2.6). If unregulated, osmosis would quickly result in fatal hemodilution in freshwater fish and hemoconcentration (dehydration) in marine species.

The environmental concentration gradients within which osmoregulation can be successful are species specific. Most fish have quite narrow tolerance limits (stenohaline) and cannot survive salinities much different from those in which they normally live. In contrast, a certain number of fishes, such as the striped bass and milkfish, can maintain a relatively constant blood osmolarity over a wide range of ambient salinities (termed *euryhaline*). For example, the commercially cultured milkfish can osmoregulate over salinity ranges of 0–100 parts per thousand (ppt) or more. However, considerable amounts of caloric energy from the diet must be used for osmoregulation when these fish are reared under high salinity conditions. Spontaneous swimming activity in milkfish is greatly reduced at

Freshwater Fish

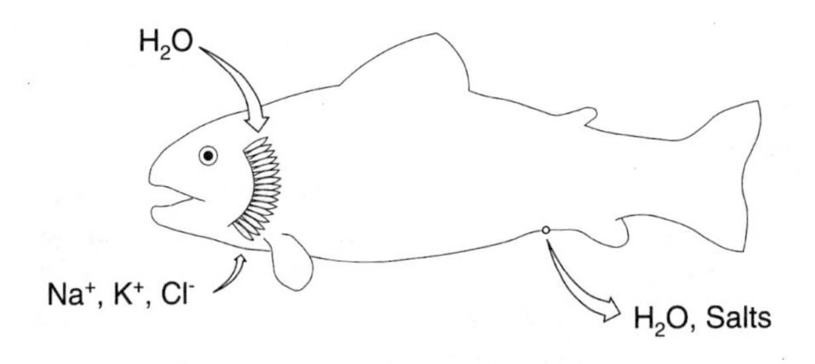

Seawater Fish

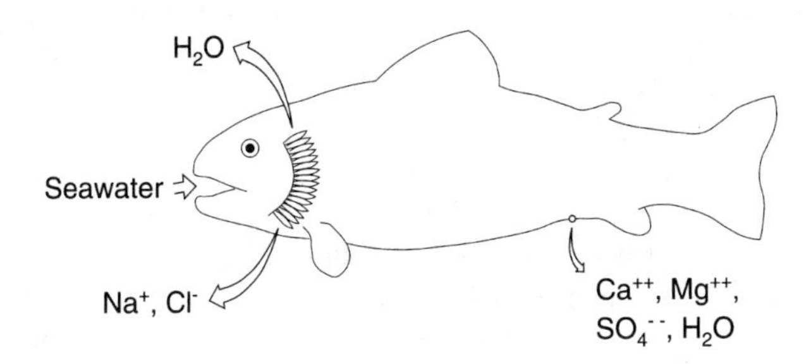

Figure 2.6. Major features of osmoregulation in freshwater and saltwater fishes (after Wedemeyer et al. 1976)

salinities over about 50 ppt reflecting the energy drain (caloric cost) of osmoregulation in such strongly hyperosmotic environments. A smaller number of fishes are anadromous and undergo a parr–smolt transformation during their freshwater rearing period that allows them to osmoregulate in the rapidly changing salinities in estuarine areas during their seaward migration as smolts, in the ocean as adults, and in freshwater again during their return spawning migration.

The main area of influx of osmotic water is the gills because they represent the largest permeable surface on a fish—from 2 to 10 cm^2/g of body weight in sedentary and active fish, respectively (Smith 1982). Changing the respiratory surface area of the gills is one form of osmoregulation. If a fish is stressed, the adrenaline and other catecholamine hormones produced dilate the branchial vasculature. This produces a functional increase in respiratory surface area and increases oxygen uptake. However, it also increases the osmoregulatory load because an increased influx of water also occurs. The balance required between the need for oxygen and the need to minimize osmosis is sometimes termed the *osmorespira-*

tory compromise. In fish culture, when hatchery salmonids are anesthetized, weighed, and measured, a water influx of about 15% of their body weight can occur during the next hour or so because of the increased oxygen consumption that occurs in response to the excitement, stress, and the temporary anoxia from being held out of the water. Similar osmotic water influxes occur in salmonids suffering from infectious hematopoietic necrosis (IHN) virus infections, although the osmoregulatory problem in this case is probably the result of gradual kidney failure, rather than the result of adrenaline and other catecholamine hormone production due to stress (Amend and Smith 1974). Fish in seawater stressed by handling or other fish cultural procedure experience weight loss.

The structure of fish skin has important implications for the effects of rearing conditions on fish health and condition. Unlike terrestrial vertebrates in which the outer layer of the epidermis is composed of dead or keratinized squamous epithelial cells, the skin of most fishes consists of living cells. Strength is provided by the stratum compactum, a noncellular layer composed primarily of collagen (Yasutake and Wales 1983). The living skin layer is easily abraded by handling and infections and osmoregulatory problems can occur very quickly. Salmonids have a relatively impermeable skin, protected by a mucoid slime layer. Fish cultural procedures involving handling, external parasite infestations, or injuries can easily disrupt the mucous layer, abrade the skin, and result in an increase in the normal osmoregulatory load. Mucus also lubricates the scales as they slide over each other during swimming, reduces frictional drag of the water, contains antibodies, and functions as a significant antibacterial agent. Secondary fungus infections almost always follow disruption of the mucous layer.

Ionoregulation is an important aspect of osmoregulation in both marine and freshwater fishes. Mitochondria–rich cells clustered around the bases of the gill filaments, termed *chloride cells*, actively transport sodium and chloride ions from the water into the blood of freshwater fish. The chloride cells help maintain homeostasis by transporting monovalent ions from the water into the blood to replace those lost by diffusion through the gills and in the copious urine that is produced. The lower the mineral content of the water, the larger the diffusion gradient between blood and water becomes. The significance of this to aquaculture is that in very soft water (<20 mg/L total hardness, as $CaCO_3$), concentration gradients of up to 3,000-fold can occur and the active transport of these ions can consume several percent of the caloric energy provided by the diet. In hard water, the energy costs are lower because the concentration gradient is lower. In seawater the concentration gradient is reversed because the environmental salt concentration is normally 3–3.5% (30–35 parts per thousand salinity) and ions tend to diffuse through the gills into the blood. The direction of active transport thus becomes the reverse of the freshwater situation.

The gill epithelium is also the site of coupled Na^+/H^+ and Cl^-/HCO_{-3} exchange between the blood and the water. This function helps regulate the acid–base balance of the fish and facilitates the excretion of ammonia and carbon dioxide. One of the adverse effects of heavy metal exposure is to inhibit these important processes. The resulting ionoregulatory dysfunction and blood ammonia increases can seriously debilitate fish health and condition. In marine fishes, all these ions move outward and an exchange system in the gills has not been documented.

The involvement of the gill ion exchange mechanism in the excretion of blood ammonia by the exchange of Na^+ from the water has implications for the health of fish reared in water of naturally low mineral content. As one example, mineral salt supplements to increase the dissolved Na^+ to 20 mg/L have been successfully used to lower blood ammonia levels in smolting steelhead trout and prevent the swollen gills and mortalities thought to be due to ambient Na^+ concentrations that were too low to allow sufficient ammonia excretion by the branchial Na^+/NH_4^+ ion exchange mechanism (Bradley and Rourke 1984).

As mentioned, the water which continuously diffuses by osmosis through the gills of freshwater fishes must also be continuously excreted by the kidneys at the same rate. Urine production is thus relatively large, as much as 1% of the body weight per hour. The implication for fish health, however, is not that large volumes of water are excreted, but that the urine begins as an ultrafiltrate of blood plasma having the same concentration of electrolytes as the blood. Extensive salt reabsorption occurs in the kidney tubules, but in spite of this, small amounts of blood electrolytes are steadily lost because the volume of the urine is so large. These electrolytes must be continuously replaced, either from the water by the gill chloride cells, or from the diet, or life–threatening ionoregulatory disturbances will quickly occur.

In marine (bony) fishes, osmotic forces cause water to move out through the gills, dehydrating the fish, while salts diffuse inward driven by the steep concentration gradient between seawater and the blood. To prevent dehydration, marine fish must drink relatively large amounts of seawater—10% or more of their body weight per day—and then physiologically distill it to remove the salts it contains. For piscivorous species, eating other fish provides a minor additional source of water.

After marine fish drink seawater, most of the ingested water and monovalent ions (sodium, potassium, chloride) pass into the blood through the gut epithelium. These ions are later excreted by the gills. Most of the divalent and trivalent ions (magnesium, calcium, sulfate, phosphate) from the seawater remain in the gut and are excreted rectally with the feces. Small quantities of these ions do pass through the intestinal epithelium into the blood and are excreted by the kidneys. However, while kidney function must continue to some degree, urination represents a loss of water that must be made up from some other source. Urine production by marine fishes is greatly reduced compared with freshwater species and the gut epithelium serves as an osmoregulatory organ as well as for nutrient absorption.

Anadromous salmonids reared in seawater net pens also minimize use of the kidney except for excretion of larger organic molecules that cannot diffuse across the gills. Urine production is typically reduced to less than 10% of the amount produced during freshwater rearing but is generally somewhat higher than strictly marine species. As with strictly marine species, the intestine in anadromous salmonids also functions in osmoregulation. In returning adult Pacific salmon for example, most of the gut atrophies to a small fraction of its original diameter after the fish stop feeding prior to entering freshwater streams to spawn. However, the posterior third of the gut remains normal in appearance and richly vascularized suggesting the continuance of normal osmoregulatory function in that area. In both marine and anadromous fishes, the caloric cost of osmoregulation in seawater can be up to 30% of the resting metabolic rate. Thus, feeding artificial diets high in moisture content may be somewhat helpful to salmonids grown in seawater net pens.

PARR–SMOLT TRANSFORMATION

A series of developmental changes occur in juvenile anadromous salmonids while they are rearing in freshwater streams and lakes that enable them to migrate to the ocean, enter seawater, and osmoregulate successfully in the strongly hyperosmotic marine environment. Collectively, these changes are variously referred to as the parr–smolt transformation, smolt development, or simply smoltification. The process of smoltification requires the initiation and coordination of numerous physiological, morphological, and behavioral changes including thyroid and growth hormone production, stimulation of gill adenosinetriphosphatase enzyme activity (ATPase), and development of a characteristic silver coloration, migratory behavior and preference for seawater. The result is the transformation of an often territorial freshwater dwelling juvenile fish (parr) into a loosely schooling ocean dwelling pelagic fish able to continue to grow and develop into an adult and then return to its freshwater natal stream to spawn. These developmental changes occur in response to natural environmental cues and can easily be adversely affected by mismanaged rearing conditions in intensive culture systems (Iwama 1992).

It is particularly important to understand the effects of rearing conditions on the physiological development of hatchery-produced Atlantic and Pacific salmon and steelhead, because these species are produced both commercially as food fish and for conservation purposes to supplement declining wild populations. Commercially produced Pacific (coho, chinook) and Atlantic salmon are normally reared to the smolt stage in shore–based freshwater hatcheries and then transferred directly into seawater net pens for growout to market size. Anadromous salmonids produced by conservation hatcheries are usually intended to supplement declining wild populations and are released into rivers to migrate naturally into the ocean. The hatchery juveniles must be physiologically and behaviorally similar to the wild fish so that they migrate to the ocean, grow and develop, and return to their natal river (hatchery) at similar rates. In both commercial and conservation situations, developmentally immature fish will not be fully prepared to osmoregulate against the threefold or higher ionic gradients of the ocean, and will either fail to survive seawater entry or will survive but fail to grow and develop normally.

A few species of Pacific salmon (chum, pink, and fall–run chinook) migrate to the ocean relatively soon after hatching and their attainment of seawater tolerance is associated mostly with reaching a certain minimum size (e.g., 5–6 grams for fall–run chinook). The Atlantic salmon, anadromous rainbow trout (steelhead), and the other Pacific salmon (coho, spring [stream type] chinook, sockeye, masu) have an extended rearing period in freshwater and undergo a more classic smolt development. The onset of smoltification in the latter group is initiated and coordinated by the normal seasonal changes in photoperiod and temperature. In turn, the development of the succeeding endogenous processes (body silvering, fin margin darkening, salinity tolerance and preference, growth rate, and gill ATPase activity) can be activated, postponed, or advanced by photoperiod manipulation. Water temperature acts as an environmental controlling factor setting the range within which these developmental changes can proceed and determining their relative rates.

A summary of the physiological changes occurring during smolt development of particular interest to fish culturists is given in Table 2.3. Of these changes, the

TABLE 2.3. Physiological changes of particular interest to fish culturists that occur during the smolt development of Atlantic and Pacific salmon and anadromous rainbow trout.

Physiological Characteristic	Change During Smolt Development
Migratory behavior	Increases
Salinity tolerance	Increases
Salinity preference	Increases
Body silvering, fin margin blackening	Increases
Growth rate	Increases
Condition factor (weight per unit length)	Decreases
Oxygen consumption	Increases
Ammonia production	Increases
Liver glycogen	Decreases
Blood glucose	Decreases
Growth hormone	Increases
Corticosteroid hormones (cortisol)	Increases
Thyroid hormone (Thyroxine)	Increases
Prolactin (hormone)	Decreases
Gill ATPase activity	Increases

Source: Wedemeyer et al. (1980).

external body silvering and fin margin blackening (in Atlantic salmon, coho salmon, and steelhead trout) are among the most dramatic characteristics and have long been used to visually distinguish smolts from parr. The silvery color results from deposition of guanine and hypoxanthine crystals in the skin and scales. These are purine compounds produced as a result of the seasonal thyroid hormone–induced increase in protein catabolism. Beef thyroid gland added to the diet or thyroid hormones such as thyroxine added to the water will cause body silvering of fish that are still parr. Elevated blood cortisol levels will also cause silvering. Thus, coloration by itself is not an adequate criterion for smolting, especially in hatchery fish. These colors can sometimes develop in juveniles that are not functionally smolts. However, silvering and fin margin blackening can be a fairly reliable indication of smolt status in wild fish.

Another physiologically dramatic aspect of the parr–smolt transformation is the development of hypoosmoregulatory ability that preadapts the fish to life in seawater by conferring salinity tolerance and preference. Again, these developmental changes occur during freshwater rearing and include a decrease in plasma and tissue chloride concentration, a decrease in glomerular filtration rate, and a dramatic increase in gill ATPase activity. However, salinity tolerance, as measured by short-term seawater survival, cannot by itself be used to distinguish true smolts from parr. For example, juvenile steelhead trout, coho salmon, and Atlantic salmon develop seawater tolerance before the remainder of the smolting process, such as development of migratory behavior, is complete. Nevertheless, the hypoosmoregulatory ability of incompletely smolted juveniles can be distinguished physiologically from that of completely developed smolts by a high salinity (40 ppt) challenge. In Atlantic salmon, tolerance to this degree of salinity does not develop before smoltification is complete, and partially smolted juveniles will not survive.

The salinity tolerance and preference of juvenile anadromous salmonids also shows a strong seasonal increase, and peak hypoosmotic regulatory capability generally coincides with the development of migratory behavior. Salinity tolerance is also influenced by fish size. In Pacific and Atlantic salmon and steelhead trout, hypoosmoregulatory capacity increases as the fish grows until smolting occurs. In nonsmolting salmonids such as the nonanadromous rainbow trout, continued growth appears to confer increased ability to tolerate seawater. Rainbow trout 6–8 in. in size or larger can generally survive gradual acclimation to full strength seawater. Chum and pink salmon can be directly transferred to seawater shortly after swim–up but will suffer a 5–10% mortality. Growth and survival are better if these species are held at salinities of 10–15 ppt or less until they attain a fish size of at least 1 gram. Sockeye salmon can be transferred to seawater at about 2 grams with low mortality if they are first acclimated to a salinity of 10–15 ppt for a few days. Chinook and coho can survive direct transfer to seawater at fish sizes of 5–6 grams and 10–15 grams respectively. If smolts are held in freshwater and not allowed to migrate to the ocean, parr-reversion will begin and their ability to adapt to seawater will decrease. However, they will continue to grow, although at a slower rate than if they were in seawater. Naturally landlocked populations of wild anadromous salmonids may show the same seasonal development of seawater tolerance as anadromous strains and can also grow as rapidly as anadromous strains of the same species if food availability is high (Foote et al. 1994).

Anadromous salmonids rearing in streams far from the ocean may begin migration before the smolts are physiologically ready to tolerate seawater. For anadromous salmonids of commercial importance to aquaculture, full smolt development is usually required before transfer to seawater net pens. In Atlantic salmon, suppressed feeding behavior and reduced growth typically occur if juveniles are transferred to seawater before smolt development has been completed. Similarly, incompletely smolted juvenile coho salmon suffer dramatic growth suppression (stunting) a few weeks after seawater entry and show a number of endocrine abnormalities as well (Clarke 1992). One widely used method to determine whether juvenile salmon have completed the process of smolt development is the seawater challenge/blood sodium test (Blackburn and Clarke 1987).

Two other indicators that fish culturists can easily use to monitor smolt development are growth rate and condition factor. The normal growth rate of both Atlantic and Pacific salmon has long been known to increase sharply during smolt development. Physiologically, it is now recognized that the period of rapid growth is an integral part of smolt development. Conversely, slow growth tends to reduce smolting tendency. If parr are injected with mammalian or teleost growth hormone, both their growth rate and salinity tolerance will increase substantially. Similarly, injections of teleost or mammalian growth hormone will improve the hypoosmoregulatory performance of nonanadromous trout to the point that they become able to survive in seawater. Hypertrophy and hyperplasia of the pituitary growth hormone cells can be seen histologically during the normal development of smoltification. Increasing the photoperiod during winter with supplemental lighting will also increase the number and granulation of pituitary growth hormone cells, promoting growth and accelerating the development of smolt characteristics (Hoar 1988).

The use of condition factor to monitor smolt status is based on the fact that the length–weight relationship normally decreases during smoltification. That is, smolts weigh less per unit length than do parr. Condition factor can usually be calculated from the normal hatchery records and thus has a long history of use in the intensive culture of steelhead trout and Atlantic and Pacific salmon.

Reduction in total body lipid stores is another physiological change that can be used to indicate stage of smolt development, although routine monitoring is rarely done in either conservation hatcheries or commercial aquaculture operations. This metabolic change is due to accelerated lipid catabolism and is related to other characteristic alterations such as elevated oxygen consumption, increased growth rate, proliferation of pituitary growth hormone cells, and changes in muscle tissue water content. The effect can be experimentally demonstrated by inducing salmon to smolt during winter by using lights to artificially increase the day length. Total lipid content will be significantly lower than in the nonsmolted controls held under a natural photoperiod. Photoperiod manipulation used to retard smoltification results in higher lipid content and lower standard oxygen consumption rates, compared with fish of similar age under natural photoperiods.

Carbohydrate metabolism is also altered during smolt development. Reduced liver glycogen and elevated blood glucose concentrations occur, together with the accelerated protein and lipid catabolism discussed above. At the same time of year, parr typically have higher liver glycogen levels and lower blood glucose than smolts. Parr also survive handling and transportation with fewer adverse effects on carbohydrate metabolism. Smolting coho salmon subjected to a similar amount of handling may suffer a life-threatening hypochloremia. In fish with subclinical bacterial kidney disease, activation of the infection is also likely.

A number of endocrine changes occur during smolt development including a surge in circulating levels of insulin, thyroxine and cortisol, as well as growth hormone. In contrast, prolactin levels gradually decrease (Dickhoff et al. 1990). In freshwater fish, prolactin favors osmoregulation by inhibiting ion excretion and stimulating ion uptake from the water. The activity of pituitary prolactin cells begins to decrease during the early stages of smolt development and blood levels of prolactin fall still further after the fish enter seawater. Growth hormone enhances salinity tolerance by stimulating both growth itself and the activity of the gill chloride cells involved in seawater osmoregulation. The surge in insulin production lowers blood glucose levels and promotes the storage of lipids and proteins. Thyroxine apparently acts to regulate the overall rate of smolt development. The pituitary-interrenal axis is also activated during smolt development leading to increased blood cortisol concentrations (Hoar 1988). Higher cortisol levels apparently favor development of seawater osmoregulatory ability but probably also temporarily affect the immune system (discussed in Chapter 5), which may account for the decreased resistance to infections that seems to accompany smolt development. A diagrammatic summary of these changes is given in Figure 2.7.

The increase in gill ATPase activity is an aspect of smolt development particularly relevant to monitoring the physiological condition of anadromous salmonids reared in intensive culture systems. This enzyme system provides the chemical energy needed by the chloride cells to actively transport monovalent ions absorbed from seawater back into the ocean. Euryhaline fishes typically show increased gill ATPase activity after they have been in seawater for a period of time. In anadromous fishes, the increase in ATPase activity begins during freshwater rearing and peaks near the time smolts show migratory behavior in a hatchery or their most active migratory behavior in a stream Figure 2.8). After the smolts enter seawater, gill ATPase activity usually continues to rise for a few weeks more and then stabi-

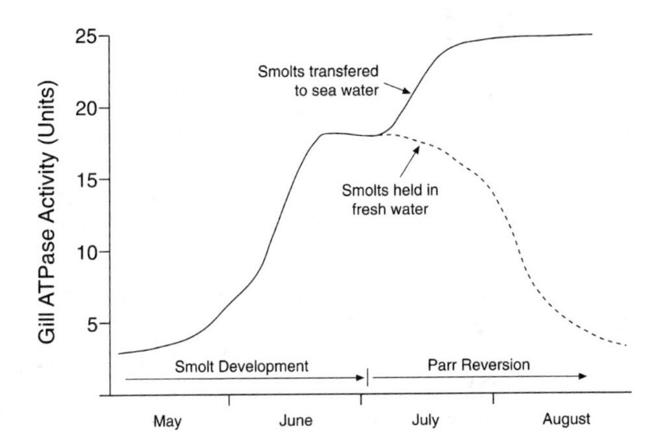

Figure 2.7. Diagrammatic summary of the principal endocrine and related changes occurring during the normal smolt development of anadromous Pacific salmon and steelhead.

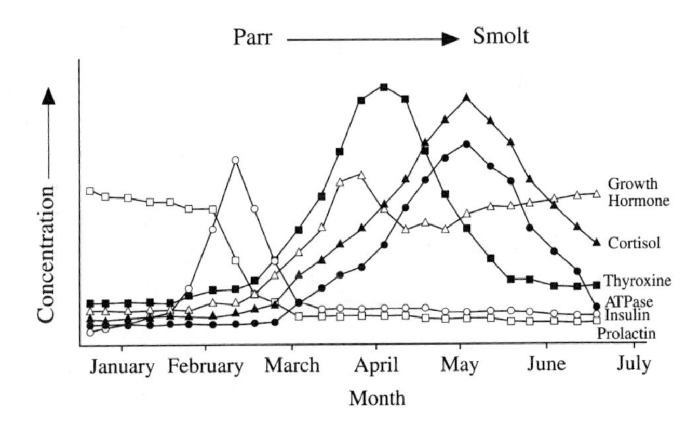

Figure 2.8. Seasonal development of gill ATPase activity in juvenile Pacific and Atlantic salmon. After seawater entry, ATPase activity continues to increase for a time. If the smolts are held in freshwater, gill ATPase activity will revert to baseline levels as parr reversion occurs (after Dickhoff et al. 1990).

lizes at the higher level. If the smolts are kept in freshwater, gill ATPase activity will gradually decline to levels typical of parr as desmoltification (parr-reversion) occurs.

The development of migratory behavior is perhaps the most dramatic behavioral change shown by smolts. Migratory behavior is easily observed in hatchery raceways and has long been used by fish culturists to assess the progress of smolt development. However, as with the other criteria, it is better used to evaluate the extent of smolt development in wild rather than hatchery fish.

Other significant behavioral changes also occur during smolt development that can be monitored by fish culturists. For example, Atlantic salmon parr are territorial and semidemersal living on or near the bottom substrate. As smolt development proceeds, the fish begin to develop schooling behavior and swim at mid-depth. The decline in territoriality may be one of the factors involved in the eventual change to a pelagic existence. However, the physiological mechanisms driving such behavioral changes are not understood. One explanation may be that smolts maintain larger volumes of air in their swim bladders than do parr. The increased buoyancy could result in more difficulty in maintaining their position on the stream bottom thus encouraging their migration downstream. Another behavioral change during smoltification is the development of a preference for higher salinities. The physiological basis for this change is also not understood. However, salinity preference can be altered by photoperiod manipulation.

Another aspect of the parr–smolt transformation important to hatchery production is the influence of fish size. Most facets of smolt development including buoyancy, silvering, and salinity tolerance proceed faster and to a greater extent in larger than in smaller parr. However, size alone is an insufficient criteria for separating smolting from nonsmolting juveniles; especially when elevated temperatures have been used to accelerate growth. For example, the minimum size threshold for smolting in Atlantic salmon appears to be about 10 cm but most smolts are in the 14–17 cm range. Both Pacific and Atlantic salmon parr rearing in rivers at the northern end of their range often grow slowly because ice often persists well into the summer and water temperatures remain low. These fish can show signs of smolt development in the spring of several successive years before they finally migrate. In some cases, Atlantic salmon smolts can be 6 years old and 20 cm or longer before they finally migrate. Smaller fish can be silvery, have a certain degree of salinity tolerance, and show increased buoyancy, but the other aspects of smolt development, especially migratory behavior, will not begin until the threshold size has been reached.

Following successful smoltification, juvenile salmonids become physiologically capable of tolerating direct transfer to seawater without an acclimation period. Some water and ion fluctuations occur, but typically all physiological systems will be functioning normally within a few days (Figure 2.9). However, even though direct seawater transfer is normally well tolerated, some stress is involved. Additional stress imposed during this time by fish cultural procedures may be sufficient to activate latent infections with fatal results. For example, coho salmon smolts with subclinical furunculosis or bacterial kidney disease infections may have the disease activated upon transfer to saltwater. *Vibrio* infections following the stress of seawater transfer are also fairly common occurrences. Smolts entering seawater with *Ichthyobodo* infestations can suffer delayed mortalities of 60–70% from osmoregulatory failure because of the physical damage to the skin and gill chloride cells caused by this protozoan parasite (Urawa 1993). Prerelease

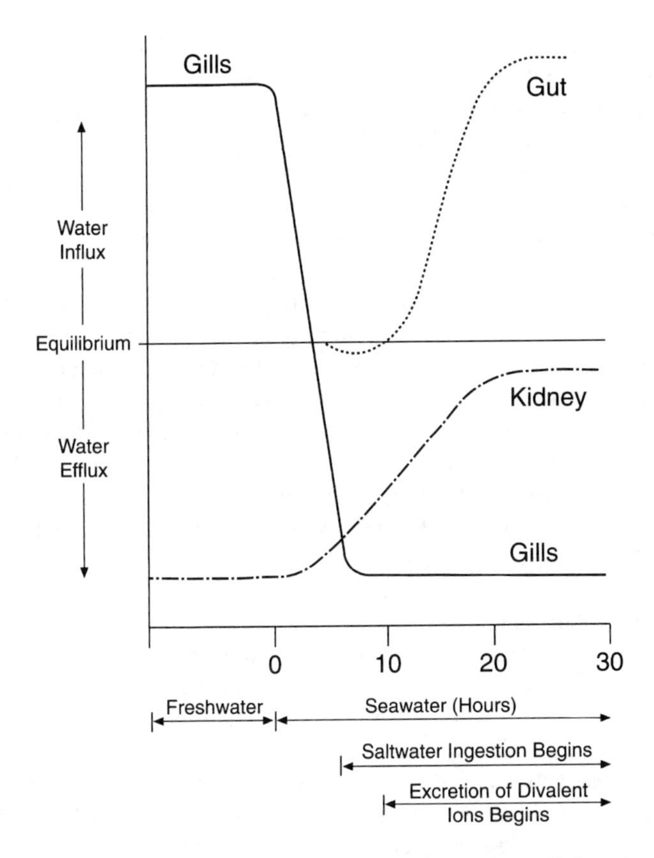

Figure 2.9. Diagrammatic representation of the changes in hydromineral balance in smolts leaving freshwater and entering seawater. Physiological systems are functioning normally within a few days (after Wedemeyer et al. 1976).

drug and chemical treatments to control these diseases are standard practice. Unfortunately, some of the drugs and chemicals used can also impair seawater tolerance and lead to reduced early marine survival if an adequate recovery time is not allowed. This subject is discussed more fully in Chapter 4.

Mismanaged rearing conditions in intensive culture systems can easily impair the parr–smolt transformation without exceeding physiological tolerance limits for normal freshwater growth or survival. Improperly managed fish densities, water temperature, water chemistry, or photoperiod are particularly liable to inhibit one or more aspects of the parr–smolt transformation without exceeding physiological tolerance limits for other life processes. For example, otherwise sublethal exposures to copper, arsenic, and other heavy metals in freshwater can partly or completely inactivate gill ATPase function, resulting in impaired migratory behavior, reduced seawater tolerance, and thus reduced early marine survival. Herbicides used in forest, agriculture, and range management can drift into streams used by juvenile salmonids and have the same effect. Acid precipitation can also impair the parr–smolt transformation of juvenile salmon by either the direct effect of low pH on gill ATPase development and growth rate or by solubilizing toxic heavy metals. Again, these effects may remain sublethal while the smolts are in freshwater, but can result in high mortality after seawater entry (Saunders et al. 1983).

Elevated water temperatures used to accelerate growth are well known for their potential to adversely affect smoltification. The normal development of gill ATPase activity in steelhead trout is suppressed at water temperatures above 13°C, resulting in "smolts" with impaired ability to osmoregulate in seawater. Atlantic salmon should also be transferred to seawater before the freshwater temperature reaches 13°C (Saunders and Dustin 1992). Water temperatures above 13°C can be used to accelerate growth if the smolts are given an adequate period of time in cooler water prior to release. Elevated water temperatures accelerate both the smolt development and subsequent parr-reversion of coho salmon thus shortening the optimum time period for release.

The chronic sublethal stress associated with intensive rearing conditions has been suggested as a reason for the impaired ocean survival of hatchery-reared juvenile salmonids as compared to naturally produced smolts. Downregulation induced by the low level chronic stresses associated with hatchery rearing conditions may be responsible for the reduced number of gill corticosteroid receptors, gill ATPase activity, and smaller surge in plasma cortisol concentrations that have been noted in hatchery fish (Shrimpton and Randall 1992). More information on preventing the adverse effects of rearing conditions that can limit the efficiency and effectiveness of intensive culture systems producing juvenile salmon for commercial aquaculture or to supplement declining wild runs will be given in Chapter 4.

FEEDING, DIGESTION, EXCRETION

Feeding

Under extensive rearing conditions, naturally produced food organisms are available and fish are not completely dependent on artificial feeds. In the highly modified rearing environments typical of intensive culture systems such as seawater net pens, suspended cages, and raceway rearing systems, fish are usually completely dependent on artificial feeds provided by the fish culturist for their nutrient supply. Although fish such as salmonids can successfully utilize commercially formulated artificial diets from the time of first feeding, the larval stages of other fishes may have behavioral or other requirements for live food. A working knowledge of the physiology of feeding and digestion is important to the development and efficient use of artificial diets that will meet both nutritional needs and behavioral feeding requirements.

The fundamental anatomical processes involved in feeding are similar in many ways in all the higher animals. However, there are also some significant differences in feeding (and digestion) related to life underwater that may be helpful to consider. First, unlike terrestrial animals living in the atmosphere, some of the minerals fish require (such as sodium, calcium, and chloride) are obtained from the water (by active transport into the blood by the gills) as well as through the diet. Second, the feed is suspended in their respiratory medium and decomposing food can quickly deplete DO and produce toxic CO_2 and ammonia. Food items should either float, or sink at a rate that will allow the fish to strike and consume them before they reach the bottom.

For fish to grow normally, food must be ingested. However, ingestion underwater is not a straightforward process. Food items have to be actively taken in, usually with a mouth full of water. Many species draw suspended food items into the mouth by rapidly increasing the size of the buccal cavity to create suction. If high water velocities are required, the mouth opening must be made quite small and a tubular, sucking mouth has evolved in some species for this purpose. Fish that strike at their prey or are scavengers generally have large mouths with wide gapes and large buccal cavities. There are a great number of intergradations between these two major types. Whatever its specialization for feeding, the mouth is usually also required to function as a respiratory pump to move water over the gills and be capable of clearing accumulated debris by forcefully reversing the direction of flow—the so-called coughing reflex.

After food enters the mouth, it is usually not processed by chewing but simply swallowed. Teeth are used primarily to manipulate prey items after capture or to handle specialized types of food. Teeth occur in a variety of places inside the mouth such as around the edge of the jaws, the anterior side of the gill bars, the soft palate, or supported in a pharyngeal muscle mass just anterior to the esophagus. Pharyngeal teeth are used by the grass carp to grind cylindrical stems of plants that are introduced lengthwise into the mouth.

Digestion

Digestion in fish begins in the stomach instead of the mouth. As in mammals, the distention produced by food entering the stomach stimulates secretion of hydrochloric acid and pepsinogen, which is then hydrolyzed to pepsin. Proteins are broken down to polypeptides and minerals are solubilized but no fat or carbohydrate digestion occurs. In marine fish, the food is first coated with a mucous layer containing hydrochloric acid and pepsin. This helps prevent dilution by the alkaline seawater which marine fish must continually drink to maintain their osmotic balance. The partially digested food mixed with mucus and gastric juices is slowly released through the pyloric sphincter into the alkaline intestine where it stimulates the release of bile and digestive enzymes such as chitinase, amylase, and lipase from the liver and pancreas. The presence or absence of swallowed seawater makes no further difference and the remainder of the digestion process is similar in both marine and freshwater fishes. Some enzymes are also secreted by the gut wall. Digested nutrients are absorbed into the blood by a combination of passive diffusion, active transport, and pinocytosis, and carried into the liver through the portal vein. An undigestible residue of 10–20% usually remains to be excreted as feces. The general process of digestion is similar in most fish important to aquaculture except in carp, which have no separate stomach and the entire digestive tract is alkaline, and in tilapia, which have a modified stomach. Fish in general do not have a separate large intestine.

An aspect of digestion of considerable importance to intensive fish culture is that oxygen consumption increases by 50% or more and intestinal blood flow may double immediately after digestion begins. The increased oxygen consumption is associated both with digestion and the metabolism of amino acids and other nutrients. In salmonids and catfish, the energy cost of these metabolic activities averages 10–15% of the calories ingested (Adams and Breck 1990). The anorexia shown by many fish as the water temperature increases or the DO decreases is probably a physiological adaptation to this energy cost. For example, at temperatures above 27°C, salmonids must consume oxygen at close to the maximum sustainable rate just to support their basal metabolism and could not survive the additional oxygen demand associated with digestion. Increased oxygen consumption also occurs as a result of stress and anorexia may also protect fish against this as well. Fasting for 24 hours prior to handling or transportation can help prevent the combined increase in oxygen consumption resulting from digestion and stress from exceeding the sustainable respiratory capabilities of the fish. The reduced production of fecal material and the toxic excretory products ammonia and carbon dioxide is an additional benefit.

Although 24 hours of fasting is a useful general rule, the actual time required for food to be digested, usually measured as the time for an inert marker to appear in the feces, is highly dependent on water temperature and on the amount and composition of the diet. Food retention time decreases as the amount of food ingested and the water temperature increase. Thus, a ration equivalent to 0.5% of the body weight might take more than several days to pass through the gut of a salmonid at 5°C while at 20°C, most of it would pass through in several hours.

All food consumed by fish is eventually utilized for metabolic processes, synthesized into new tissue (growth), or excreted as waste. The proportion of ingested food that is absorbed by the fish is also temperature dependent. In salmonids, about 80% of a typical formulated ration can be absorbed and become available as net energy at low water temperatures (e.g., 5°C). At 20°C, only about 60% may be absorbed. Proximate analysis of the natural food organisms of salmonids and channel catfish has led to the development of artificial diets with 35–40% (or more) protein and less than 12% carbohydrate (dry weight basis) with the remainder consisting of lipids, vitamins, minerals, and inert roughage. The kinds and quantities of constituents used to achieve the desired levels of protein, carbohydrates, and lipids is dictated by cost and digestibility considerations. Digestibility determines the effective caloric value of the basic protein, carbohydrate, and lipid diet constituents. For example, starch is only about 38% digestible. As a consequence, proteins and carbohydrates both contain about 4.6 kcal/g but fish can obtain only about 3.9 and 1.6 kcal/g from them, respectively. Lipids contain 9.4 kcal/g, but their effective caloric value to fish is about 8 kcal/g. Digestibility can also be reduced during processing if, as sometimes happens, ingredients such as fishmeals are overheated while being dried.

Although meeting nutritional requirements for protein, carbohydrates, and lipids is important, the efficiency of converting food into tissue (growth) is the overriding economic factor in intensive fish culture. Food conversion efficiency is basically an intrinsic function of the balance between carbohydrates, lipids, and proteins within the feed, although it can also be affected by extrinsic factors in the rearing environment. Physiologically, carbohydrates and lipids both exert a sparing action on protein; that is, if sufficient carbohydrates and lipids are available, most protein is utilized for growth rather than being metabolized to obtain energy. Because protein is normally the most expensive part of the ration, diets for salmonids and catfish are usually formulated with only 90–100 mg protein per kcal to minimize use of proteins as an energy source (Wilson and Halver 1986). Increasing the dietary carbohydrate level will spare protein for growth but levels above 10–15% can cause liver glycogen deposits that will result in eventual liver pathology. Similarly, dietary lipid levels of 15–20% accomplish protein sparing but levels above this result in visceral fat deposits that will increase costs by reducing dress out weights in fish produced for food. Saturated fats added to diets to spare protein can lead to harmful visceral lipid deposits because fish enzyme systems are evolutionarily adapted to oxidize unsaturated lipids.

The efficiency of converting food to tissue is usually expressed in terms of the feed conversion ratio, weight gained (kg)/weight of feed fed (kg). Food conversion for salmonids can be as high as 1.1:1. That is 1.1 kg of food will produce 1 kg of fish tissue. In the extensive culture of warmwater fishes, such as channel catfish, milkfish, or tilapia, the feed conversion ratio can be as favorable as 0.6:1, but this ratio is usually due to the availability of natural food in the ponds. In general, young fish grow more efficiently than older fish and the more abundant the food, the less efficient the growth. However growth efficiency is also strongly affected by water temperature. For example, if sockeye salmon are fed to satiation, the optimum temperature for growth is near 15°C. If the feeding rate is reduced to 1% of body weight/day, the optimum temperature for growth decreases to about 5°C.

Although it is widely accepted that the health and physiological condition of fishes used in aquaculture depend on good nutrition, detailed information on specific nutritional requirements is available for only the channel catfish, tilapia, and some of the salmonids. Based on this information, commercially produced moist or dry diets have been developed and are available either as floating or sinking pellets. Formulated diets used for catfish and tilapia are similar to those for the carnivorous salmonids except that they contain less protein and more carbohydrate. Unfortunately, the best diet for maximum food conversion and growth may not be the best diet for producing high quality (high survival) eggs in broodstock. The nutritional requirements of most other teleost fish have not been determined. Commercially formulated diets are available for the many species of tropical fishes reared for the aquarium trade but most of these are natural products dried into flakes or meal, which have been found on a trial and error basis to work satisfactorily for particular species.

An energy budget for fish can be constructed that relates the calories in the food consumed to calories used to provide energy for metabolism, calories used to synthesize tissue for growth and reproduction (somatic and gonadal tissue), and calories excreted as waste (feces and urine). The caloric energy from the diet available to synthesize fish tissue (growth) is the item of interest in aquaculture. Growth is largely due to protein synthesis, lipid storage, and gonad development and is usually measured as a change in weight or length. In food fish production, it is important to maximize the caloric energy devoted to growth through protein synthesis. Gonad development, in particular, consumes significant amounts of energy that would otherwise be available for growth. This is evidenced by the fact that fish can lose several percent of their body weight and a major portion of their lipid energy reserves during spawning. The caloric energy devoted to lipid storage and to the development of reproductive products instead of to growth becomes less of an important consideration if broodstock are being maintained.

Computerized physiological energy budgets (bioenergetic models) have been developed for a variety of game and forage species including the salmonids, centrarchids, and percids. In fishery management, these bioenergetics models are primarily used to estimate the food consumption or growth of wild fish populations. In fish culture, food consumption is known and the mathematical models are used to provide insight into problems such as the effects of stressful rearing conditions on growth and biomass production. For more detailed information, see the excellent review by Rice (1990) of the application of bioenergetics modeling to the evaluation of stress in fish.

The amount of energy lost through the excretion of waste products depends on diet composition and also varies with fish species. The assimilation efficiency of herbivorous fishes, whose diet is high in fiber, is generally less than 80% as measured by fecal collection. In contrast, the assimilation efficiency of piscivorous fishes such as salmonids averages greater than 90%.

Excretion

For practical purposes, all the food assimilated by animals is eventually metabolized and excreted. Even food constituents incorporated into cellular structures like bone and muscle are eventually metabolized and excreted because the components of tissue are periodically replaced. In fish, the major end products of metabolism are water, carbon dioxide, and ammonia, together with small amounts of urea, creatine, creatinine, and uric acid. Lipids and carbohydrates are metabolized directly to water and carbon dioxide. Proteins, peptides, and amino acids are deaminated to yield ammonia and the carbon chain is oxidized to carbon dioxide and water. Nucleic acids (DNA and RNA) are primarily metabolized to creatine and creatinine.

Ammonia (NH_3) is the major nitrogenous waste product excreted by fish in freshwater. Salmonids fed commercial dry diets typically produce 25–35 grams of ammonia per kg of feed consumed. As one consequence, it is easily possible for the total ammonia concentration in a fish distribution tank to reach 10 mg/L or more within a few hours. Ammonia excretion is bioenergetically more efficient than the urea and uric acid excreted by terrestrial vertebrates in that the chemical potential energy remaining in the molecule is less. Although ammonia is more toxic, large quantities of water are normally available to dilute it. Most of the ammonia is produced in the liver, converted to a nontoxic form in the blood (glutamine) and transported to the gills where it then diffuses rapidly into the water as NH_3. Some blood NH_4^+ is exchanged by active transport for Na^+ in the water. If the water pH and ammonia concentration are lower than the blood, freshwater fish can readily excrete blood ammonia. If the water pH is more alkaline than the blood and the dissolved ammonia concentration higher, the outward flow of ammonia is hindered.

The copious urine produced by freshwater fish, because of the continuous osmotic influx of water, serves to dilute the small amounts of ammonia excreted by the kidneys. In marine fishes much less urine is produced and more ammonia is excreted through the gills. A few species that are periodic air breathers, such as the mudskipper which lives in shallow coastal tide flat areas, can switch from ammonia excretion to the production and storage of urea when out of the water. When they return to the water, the urea is flushed away. Sharks and rays also produce urea, but as a way of increasing the osmolarity of their body fluids to approximate the concentration of seawater. Bony (teleost) fish lack osmoregulated urea synthesis and have evolved the previously described mechanisms for maintaining homeostasis in the face of the large concentration gradients between their body fluids and the aquatic environment.

Serious fish health problems can occur if these toxic excretory products are allowed to accumulate in intensive culture systems. Fish transport tanks and raceways with water reuse systems (biofilters for ammonia oxidation) are particularly likely problem areas.

After ammonia is excreted into the water, the hydrogen ions present convert it into an equilibrium mixture of ammonium ions (NH_4^+), which are essentially non-toxic, and un-ionized NH_3 molecules, which are toxic and can cause gill damage if they are allowed to accumulate to concentrations higher than 0.1–0.5 mg/L (discussed in Chapter 3). The amount of toxic NH_3 that will be formed is a function of the water pH, temperature, and the salinity or total dissolved solids (TDS) concentration. Because chemical analysis measures only the sum of both forms of ammonia present (i.e., NH_3 and NH_4^+), the amount of NH_3 that the fish will be exposed to must therefore be calculated. The equilibrium equation for NH_3/NH_4^+ and calculations to determine the amount of NH_3 that will be present at the pH and water temperatures of most interest in intensive fish culture operations is discussed in Chapter 3.

The other major excretory product of significance to the health and condition of fish in intensive culture is carbon dioxide. In general, salmonids produce about 1.4 mg of CO_2 for each mg of oxygen that they consume (Colt and Tchobanoglous 1981). Thus, the amount of CO_2 produced by fish consuming only average amounts of oxygen can be quite large (see Tables 2.1, 2.2). Because of its solubility, CO_2 will also accumulate rather quickly unless steps are taken to prevent it. In fish transport operations, concentrations of 20–30 mg/L within 30 minutes after loading are not unusual. In salmonids, these concentrations are sufficient to cause respiratory stress from impaired oxygen transport to the tissues due to the strong Bohr effect in these species. Under hatchery conditions most fishes can tolerate moderately elevated dissolved CO_2 concentrations if the rate of increase is slow because the blood buffering system can compensate by increasing the plasma bicarbonate concentration sufficiently to prevent acidosis. In fish transport situations, however, CO_2 will usually accumulate too rapidly for physiological compensation to occur. Millimolar quantities of sodium bicarbonate added to the hauling tank water will improve survival by offsetting the decrease in blood buffering capacity caused by hypercapnia thus maintaining the blood pH (Haswell et al. 1982).

IMMUNE PROTECTION

Because of the crowded conditions in intensive culture systems and the scarcity of water supplies free of aquatic microorganisms, a normally functioning immune system is absolutely essential to the health and physiological condition of the fish being produced. The high fish densities used tend to both increase initial susceptibility to infections and to facilitate the horizontal (fish-to-fish) transmission of pathogens when infectious disease outbreaks do occur. A basic understanding of the potentials and limits of the fish immune protection system can help in managing rearing conditions to minimize adverse effects on natural disease resistance and the necessity to rely on drug and chemical therapeutants to control the resulting epizootics. Problems with antibiotic-resistant pathogens, government restrictions, and treatment costs are increasingly making drug therapy a less desirable option.

Of the commercially important fishes presently used in aquaculture, the immune systems of salmonids and the channel catfish are the best understood. Both species employ nonspecific and specific defense mechanisms that protect against infectious agents in a manner generally similar to the immune system of the higher vertebrates (Anderson 1990). As the name implies, nonspecific protective responses are directed against microorganisms and foreign materials in general and do not target specific pathogens or antigens. Nonspecific defenses are an extremely important part of the immune protection system in fish (Anderson and Siwicki 1994). The specific protective response targets genetically determined antigens on invading microorganisms that have stimulated the fish's immune system to produce activated T-lymphocytes and antibodies.

The first line of defense against infectious agents is nonspecific physical and chemical barriers to their entry into fish tissue. The mucous coating of the skin and gills contains viscous mucopolysaccharides and bacteriolytic enzymes that trap, immobilize, and lyse bacteria of all kinds—pathogens and nonpathogens alike. Bacterial cells that survive this defense and penetrate into the tissues, perhaps through a cut or abrasion, elicit a nonspecific inflammatory response. Tissue macrophages and circulating leukocytes (neutrophils) are chemically attracted to the site and immediately begin to phagocytize and destroy the invaders. Phagocytes kill ingested bacteria by first packaging them within a phagosome. The phagosome is fused with a lysosome that releases cytotoxins and lytic enzymes that kill and disintegrate the engulfed bacteria. As mentioned, macrophages can also remove organic compounds, such as petroleum hydrocarbons and pesticides that exist as insoluble fluid droplets, by the process of pinocytosis. In the case of invading viruses, nonspecific cytokines such as interferon-1 are produced that inhibit the replication of viruses in general.

Specific components of invading pathogens, usually genetically determined parts of their physical structure (termed *antigens*), are also targeted by the immune system. For example, the protein and carbohydrate structures of bacterial cell walls and viral envelopes are often antigenic. Antigens may also be soluble, such as the proteolytic enzymes produced by bacterial pathogens to facilitate their penetration into tissues. Either of these antigen types can stimulate the fish's immune system to produce genetically programmed protein molecules (antibodies), which then bind to the antigen and inactivate it or facilitate its uptake by macrophages.

Protection afforded by the immune system against fish diseases normally involves overlapping actions by both the nonspecific and specific defense systems. Thus, if invading bacterial pathogens survive the nonspecific chemical and physical barriers in the mucous coating of the gills and skin and an infection develops, the tissue macrophages and neutrophils which are chemically attracted to the site to phagocytize the bacteria also transport some of their antigens to the melano-macrophage aggregates[2] in the spleen and kidney. Lymphocytes specific to these antigens (a subpopulation of lymphocytes termed *T cells*) then begin to proliferate. This stimulates proliferation of subpopulations of similarly specific lymphocyte B cells that can synthesize antibodies. Cytokines such as interleukin-2 are also produced that further activate the T cells, and the B cells that produce antibodies. Some of the activated B cells do not immediately produce antibodies but remain as memory cells in the lymphoid tissue where they may reside for years; ready for future encounters with the antigen(s) that initially activated them.

Within a few days, (or weeks, depending on the water temperature) the proliferating B lymphocytes begin to transport and release the newly synthesized antibodies into the circulatory system. The antibodies, which are large tetrameric polypeptide molecules, physically bind to the antigens they were programmed to recognize. If the antigen is part of a cell surface, the antibodies will link several cells together into clumps enabling macrophages to ingest several pathogens at one time. The antibody-antigen complexes that are formed also activate enzymatic proteins circulating in the blood, collectively termed *complement*, that nonspecifically lyse bacterial cell membranes, attract other immune cells, and generally amplify the whole process. If the fish survives the initial infection, the memory cells that were produced enable the immune system to remember the antigens and a faster, more intense antibody response will occur if the pathogen is encountered again—months or even years later. Vaccines take advantage of this anamnestic capability of the teleost immune system and several are now commercially available to aquaculture facilities.

Relatively straightforward techniques are available for assessing effects of rearing conditions on the immune protection systems of fish in intensive culture. Two of the tests commonly used are (1) evaluation of antibody titer in blood by a humoral agglutination response, and (2) measurement of the mitogenic and antibody-producing capacity of lymphocytes (Anderson 1990). In the first method, serum from immunized fish is used to clump antigen particles such as formalin-killed bacteria. The second method employs a modified plaque assay to quantify the amount of antibody generated by antigen-stimulated lymphocytes harvested from the hemopoietic tissue and cultured in vitro. Using these tests, it has now been shown that several of the stress factors common to intensive culture systems can impair immune functioning to the point that normal resistance to infectious diseases is decreased.

Of the rearing conditions that can adversely affect the immune system, handling and transportation, crowding, adverse water quality conditions, and nutri-

---

2. Melanin-containing cells (melanocytes and macrophages) are commonly found in fish spleen and kidneys, either scattered randomly throughout the tissue or aggregated into so-called melanomacrophage centers.

tional status are probably the best understood. For example, antibody titers decrease in lake trout held at high densities in raceways. Handling stress depresses both immune function, as shown by the passive hemolytic plaque assay, and the resistance of juvenile chinook salmon to challenge with water-borne pathogens (Anderson 1990). Collecting and transporting spring (stream type) chinook smolts suppresses the generation of lymphocyte antibody-producing cells (APC) and susceptibility to *Vibrio* infections after release increases correspondingly. The immune dysfunction is apparently a side effect of the normal hypercortisolemia that occurs during the physiological stress response. Fish injected with cortisol exhibit the same immune suppression and decreased resistance to *Vibrio* infections as fish stressed by handling. Similarly, cortisol added to fish leukocyte cultures in vitro also suppress the APC response (Maule et al. 1989).

The fish immune system is also sensitive to adverse water quality conditions. Chronic sublethal exposure to contaminants such as pesticides and heavy metals results in a wide array of immunosuppressive effects. For example, the leukocytes involved in generating immunological memory were affected in juvenile chinook salmon collected from a contaminated urban estuary (Arkoosh et al. 1991). Similarly, exposure to the insecticide Endrin impairs APC generation whereas sublethal copper or aluminum exposure impairs phagocytic activity. Other important functions of tissue macrophages, such as pinocytosis and their chemotactic response (migration toward invading bacteria), are also frequently found to be inhibited in fish taken from polluted natural waters (Weeks et al. 1989).

Immunosuppression also occurs naturally during certain stages of physiological development and thus is not always simply the result of adverse rearing conditions. In anadromous salmonids, a degree of immunosuppression occurs during the parr–smolt transformation of juvenile fish and also in returning adults during their upstream spawning migration. In smolts, both the plasma cortisol level and the number of leukocyte cortisol receptors naturally increase during this stage of the life cycle and APC production is correspondingly decreased (Maule et al. 1987). The loss of natural disease resistance shown by returning adult salmon may also be related to the decreased ability of their peripheral leukocytes to generate APC. This may be due to the rapid aging that is occurring as well as to the high blood concentrations of cortisol and steroid reproductive hormones that occur during the spawning migration. Steroids such as testosterone that are involved in sexual maturation are known to suppress the APC response when added to leukocyte cultures in vitro (Slater and Schreck 1993).

STRESS RESPONSE

In order for fish to survive, their physiological systems must be capable of adjusting to challenges from the numerous naturally occurring changes in chemical, physical, and biological conditions in the aquatic environment. These challenges can range from effects of water chemistry alterations to behavioral conflicts resulting from social dominance hierarchies. In the oceans, environmental conditions may be relatively stable but fish in estuaries and freshwater lakes and ponds can face variability that can be extreme. For fish in artificial culture, challenges to physiological systems due to fish cultural procedures are imposed in addition to those occurring naturally. For example, handling, crowding, transportation, and disease treatments are required in most hatchery operations. Other challenges, such as the rapid osmotic changes and reductions in blood oxygen faced by smolts released directly into seawater net pens, are unique to particular types of culture methods. In either situation, physiological systems must react and adjust to the altered conditions or the probability of survival will be decreased. If the altered physiological state can compensate for the challenge in question, the probability of survival will be increased but often at a cost to growth and reproduction.

A basic understanding of the physiology of the stress response, and the environmental alterations to which fish can adapt through this response, is important to identifying stressful rearing conditions and developing methods to mitigate their adverse effects on health and physiological condition. In fish transport operations, for example, the corticosteroid hormone response that occurs has been used to show that the stress of the collection and loading procedures is more severe than the process of trucking itself, both in salmonids and in warmwater fishes (Maule et al. 1988). Similar kinds of information can be used to identify and minimize the stressful effects of other aspects of intensive fish culture.

The sum of the physiological changes that occur as fish react to physical, chemical, or biological challenges and attempt to compensate is commonly referred to as the *stress response* (Wedemeyer et al. 1990). The challenges themselves, if they are severe enough to require an energy-demanding compensatory physiological response by the affected fish, are referred to as *stress factors*, or *stressors*. Challenges by acute or chronic stressors that exceed acclimation tolerance limits of the fish will reduce the probability of survival. Otherwise sublethal stress may still debilitate health if it becomes chronic because of the caloric energy required to compensate physiologically. In addition, the physiological effects of individual sublethal stressors are cumulative and survival may eventually be reduced even though separately they do not exceed acclimation tolerance limits (Barton and Iwama 1991).

The physiological changes that occur as fish attempt to compensate for stressful physical, chemical, or biological challenges they cannot avoid are similar in many ways to those occurring in the higher animals (Wedemeyer et al. 1990). In response to a stressful event such as handling or crowding, the hypothalamic portion of the brain stimulates the pituitary to release adrenocorticotropic hormone (ACTH). ACTH is circulated to the anterior kidney where it stimulates the interrenal cells to produce cortisol and other corticosteroid hormones. The sympathetic nervous system stimulates the chromaffin tissue of the anterior kidney to produce catecholamine hormones including adrenaline. Both classes of hormones initiate cardiovascular changes such as increased blood flow and pressure, increase oxygen consumption, and initiate secondary blood chemistry changes such as hyperglycemia. These physiological changes are adaptive in nature and normally enable the fish to compensate and improve its probability of survival. However, if the stress is too severe or becomes chronic, the altered state will eventually become maladaptive and physiological condition will be debilitated. For example, the increased blood flow improves gill perfusion. To the extent that gas exchange is facilitated, an increased osmotic influx of water occurs, which then must be excreted (the previously discussed osmorespiratory compromise). Because blood electrolytes are unavoidably lost in the resulting diuresis, ionoregulatory imbalances soon occur that will become life threatening if prolonged. A significant degree of immunosuppression also occurs as a side effect of the stress response and latent infections can become activated (Barton and Iwama 1991). In the case of smolts, stress may adversely affect essential performance characteristics such as migratory behavior and ability to osmoregulate in seawater—resulting in reduced early marine survival (Schreck et al. 1989). In smolts transported to seawater net pens for commercial aquaculture, impaired migratory behavior is probably of little practical significance. As mentioned however, the initial corticosteroid hormone changes (which can impair disease resistance) and the ionoregulatory dysfunctions (which can be mitigated to improve survival) are of considerable practical significance in smolts produced either for aquaculture or for conservation purposes.

The general pattern of the stress response tends to be similar whether the challenge has resulted from fish cultural procedures (netting, transportation, disease treatments), water chemistry changes (turbidity, pH, temperature), or behavioral factors (fright, dominance hierarchies). A convenient paradigm for the stress response is to think of it as occurring in three stages:

1. Initial Alarm Reaction. The pituitary-interrenal axis is activated and catecholamine and corticosteroid hormones, which initiate a series of compensatory cardiovascular and blood chemistry changes, are released.
2. Stage of Resistance. Physiological systems have successfully compensated and acclimation has been achieved. There is a caloric energy cost for compensation and growth may be reduced.
3. Stage of Exhaustion. The duration or severity of the stressful challenge has exceeded acclimation tolerance limits and the physiological changes needed to maintain homeostasis have become maladaptive. Immune protection is impaired and fish diseases may occur.

Under some conditions, fish may not show all the changes of this classical three-stage pattern. For example, exposure to toxic concentrations of cadmium does not elicit a stress response in striped mullet even when survival has been impaired (Thomas 1990). Schreck (1981) first pointed out that the classical generalized reaction to stress seems to occur in fish reacting to challenges that cause a fright response or pain (analogous to the initial alarm reaction outlined above).

Another useful conceptual framework is to consider the compensatory changes that occur as divided into primary, secondary, and tertiary responses that involve succeedingly higher levels of biological organization. If stress becomes chronic, the cumulative effects may ultimately be manifested at the population or ecosystem level.

1. Primary Response (the endocrine system). Following perception of a stressful stimulus by the central nervous system, corticotrophin releasing factor from the hypothalamus stimulates the pituitary to release adrenocorticotropic hormone (ACTH). ACTH is circulated to the interrenal cells in the anterior kidney and stimulates them to secrete cortisol. The chromaffin tissue of the anterior kidney is stimulated by the sympathetic nervous system to release adrenaline and other catecholamine hormones.
2. Secondary Response (blood and tissue alterations). Adaptive changes in blood and tissue chemistry and in hematology begin such as increased gill perfusion, elevated blood sugar (hyperglycemia), and reduced blood-clotting time. Maladaptive changes, such as lymphopenia, hemorrhagic thymus, and interrenal hypertrophy may eventually occur. Blood electrolyte losses due to diuresis may lead to life-threatening ionoregulatory failure and circulatory system collapse.
3. Tertiary Response (whole-animal changes). Reduced growth, impaired resistance to infectious diseases, and behavioral changes such as impaired reproductive behavior may lead to reduced survival.
4. Quaternary Response (populations, ecosystem). Recruitment to succeeding life stages may be decreased sufficiently to result in population declines. At the community or ecosystem level, disruptions in energy flow through trophic levels may eventually result in altered species composition.

As mentioned, the functional significance of the stress response is to improve the probability that the fish will survive physical, chemical, or biological challenges. Thus, the physiological changes are initially adaptive, that is, they help maintain homeostasis. For example, the hyperglycemia produced by the action of adrenaline and other catecholamine hormones secreted during the alarm reaction functions to provide caloric energy needed for the fight-or-flight response. The primary source of the elevated blood sugar is liver glycogen, which quickly becomes depleted. If the challenge persists, cortisol is produced and the elevated blood sugar level will be sustained by liver gluconeogenesis,[3] suppression of peripheral sugar uptake, and modification of the action of other glycemic hormones. If severe muscular exertion occurs, enough lactic acid may be formed to deplete

3. The synthesis of glucose from noncarbohydrate sources such as amino acids when the supply of carbohydrates is insufficient to meet the fish's energy needs.

the blood alkaline reserve causing metabolic acidosis and respiratory distress because of the Bohr effect.

The tolerance limits to stressful rearing conditions are genetically determined making some species innately more resistant than others. For example, brook trout are more resistant to handling and transport stress than lake trout (McDonald et al. 1993). However, all fish eventually become exhausted by stress if it is too severe and prolonged. At this point, the initially adaptive compensatory response will begin to become maladaptive, and health and physiological condition will deteriorate. For example, the corticosteroid hormones released during the primary response have both anti-inflammatory and immunosuppressive side effects. If stress becomes chronic, these effects, together with the catabolic gluconeogenic action of these corticosteroids, will be manifested as reduced growth, reduced white blood cell count, and suppressed disease resistance (Wedemeyer and Goodyear 1984). The catecholamine hormones released act to increase gill perfusion to facilitate oxygen and carbon dioxide exchange. However, the osmotic influx of water that occurs as a side effect stimulates a diuresis that can result in life-threatening blood electrolyte imbalances and circulatory collapse if prolonged. For example, salmonids held out of water while netting suffer a marked hypochloremia and if the blood chloride concentration falls below about 90 meq/L, they may not recover from the handling stress. If the stress is within physiological tolerance limits, blood chloride levels will gradually be regulated back to initial levels over a period of about 24 hours as the fish recovers.

The significance to overall fish health and condition of all of the individual stress-induced changes in physiological function is as yet unknown. However, it is prudent to assume that whole-animal performance will eventually be affected, directly or indirectly, if stress becomes chronic (Beitinger and McCauley 1990). One reason for this assumption is that physiological compensation for stressful challenges requires caloric energy. The oxygen consumption rate of stressed fish can increase by 20% or more and these calories (about 1 calorie per 3 mg O_2 consumed) thereby become unavailable for other functions that require energy such as digestion and growth.

The individual physiological changes occurring during the stress response can be used to identify rearing conditions that promote the health and physiological condition of fish in intensive culture as well as those with potential for adverse effects (Wedemeyer et al. 1990). Analyses of circulating corticosteroid and catecholamine hormone concentrations give direct information of the severity of stress and the time fish need for recovery. Immunocompetence tests, such as the lymphocyte antibody-producing cell assay, allow judgements to be made about the potential for effects on disease resistance. The severity and duration of the primary stress response can be assessed by monitoring the rise and fall of plasma corticosteroid or catecholamine hormone concentrations. Plasma cortisol measurements are used more widely than epinephrine or norepinephrine analyses, which are more difficult and expensive to conduct. True resting (control) blood concentration levels of the catecholamine hormones are also more difficult to determine.

The secondary blood and tissue chemistry changes that occur as the result of endocrine activity can also be used to characterize the severity of stress and the time fish need for recovery. There are two main advantages: The analytical methods are usually simpler and less expensive, and secondary changes provide physiological information integrated at a higher level of biological organization. The most commonly measured secondary changes are serum enzymes to detect organ and tissue damage, blood glucose as an indirect measure of stress hormone activity, and blood electrolytes such as Cl^- to evaluate effects on ionoregulation. The methods are inexpensive and a hundred or more samples can easily be run in a day. Glucose and chloride determinations can also be useful in guiding the development of stress-mitigation procedures for fish cultural procedures, such as the use of mineral salt formulations when transporting fish. The accumulation of lactic acid in muscle or blood (hyperlacticemia) is also well accepted as an indicator. Lactic acidosis occurs when excessive swimming activity has been severe enough to have caused an oxygen debt. It is also a contributing factor in the mortalities that can occur when fish are exercised to exhaustion.

Hematological determinations can also provide useful information about the tolerance of fish to an applied stress factor. For example, acclimation to temperature-induced increases in oxygen demand by salmonids is commonly accompanied by erythropoiesis and increases in blood hemoglobin. However, the changes are relatively modest by comparison with cardiovascular and ventilatory responses and probably make a minor contribution to acclimation (Houston and Schrap 1994). Changes in the blood erythrocyte count (as approximated by the hematocrit) or in hemoglobin values following acute stress can also indicate that hemodilution or hemoconcentration has occurred. However, direct measurement of plasma water is the preferred indicator of dehydration because anemias, stress polycythemia (increased red blood cell numbers), or erythrocyte swelling may occur in certain situations. Of all the hematological measurements, blood-clotting time and changes in the differential leucocyte count are among the most sensitive indicators of acute stress. Increased leucocyte counts (leucocytosis) are not normally associated with stress, but leucopenia does occur during the compensatory response. In general, stress results in lymphopenia, monocytopenia, and neutrophilia.[4] ACTH and the corticosteroid hormones are probably involved but the mechanism is incompletely understood. The end result is suppression of the immune response and increased susceptibility to infectious diseases (Wedemeyer and Goodyear 1984).

Because many of the biochemical changes that occur in response to stress are the end result of cellular pathology, histological examinations can sometimes provide information on the effect of stress factors on fishes. Secondary tissue

4. Low monocyte and lymphocyte blood counts, high neutrophil counts.

changes that can be used to evaluate the stress of rearing conditions and the time fish need for recovery include liver glycogen depletion, decreased spleen weight, gastric mucosal atrophy, and changes in normal gill and chloride cell histology. Pathological changes in tissues such as interrenal hypertrophy, atrophy of the gastric mucosa, and cellular changes in the spleen also occur in fish under stress, especially chronic stress. Interrenal hypertrophy, in particular, is pathognomonic of stress. A semiquantitative histological method to detect this condition has been developed in which nuclear diameters or cell sizes are measured in stained sections of anterior kidney tissue. Stress first causes depletion of interrenal vitamin C, corticosteroids, and other lipids. Under chronic stress, interrenal hypertrophy occurs, followed by eventual cellular degeneration and the irreversible loss of tissue function. Changes in the gastric mucosa, including decreased mucous cell diameter, have been used as indices of stress in eel culture. Subordinate animals in the social hierarchies that develop among cultured European eels show measurable degenerative changes in the gastric mucosa within 5–10 days. Atrophy of the gastric mucosa is evaluated histologically as decreased mucous cell diameter (Peters et al. 1981).

Other cellular and tissue chemistry changes that have been used as indicators of adverse environmental conditions include changes in skin mucus, gill sialic acid content, muscle or liver adenylate energy charge, muscle and liver glycogen, interrenal vitamin C, relative liver weight, and mechanical properties of bone (Mayer et al. 1992).

Several whole-animal changes in performance capacity (tertiary response level) that result from adverse effects at lower levels of biological organization can be used as indicators that rearing conditions have become stressful. Holistic responses that can be used to indicate that the limits of acclimation are being reached include changes in disease resistance, metabolic rate, growth, and reproductive success, and tolerance to standardized challenge tests (Barton and Iwama 1991).

Reduced disease resistance resulting from stressful rearing conditions is of particular importance to aquaculture because of the high economic costs of epizootics. In freshwater fishes, aeromonad, pseudomonad, and flavobacterium infections serve as very useful indicators that environmental conditions have been allowed to become marginal because these pathogens are continuously present in untreated surface waters. A more complete discussion of stress-mediated fish diseases in intensive culture systems is given in Chapter 5.

Tolerance to standardized stressor challenge tests as a method of evaluating effects of rearing conditions on health and physiological condition have considerable promise. Recommended methods can be found in Wedemeyer et al. (1990). Protocols specific to salmonids have been developed for salinity (30 ppt), pH (9.4 and 3.5), thermal (1°C/h), handling (30 sec dip net), and other stressful challenges (Iwama et al. 1992).

Changes in metabolic rate as an index of performance capacity affected by stress can be estimated by measuring oxygen consumption—usually in a respirometer. This provides a basis for estimating the effects of stress on bioenergetic budget and thus on scope for growth and activity. By measuring the caloric energy absorbed from food (A), and the energy lost via respiration (R) and excretion (U), the energy remaining for growth and reproduction (P) can be calculated from:

$$P = A - (R + U)$$

Oxygen consumption, nitrogen excretion, amount of food ingested, and the gut absorption efficiency are measured and converted to equivalent units of energy (joules per hour).

A stress-induced increase in metabolic rate consumes energy within the fish's metabolic scope for activity that could have been used for some other function such as digestion. Barton and Schreck (1987) have shown that in salmonids, as much as 25% of the scope for activity may be required to cope with even mild and brief disturbances; the fraction is likely higher for more severe stress. Respirometry can be difficult because it is quite sensitive to metabolic changes caused by brief physical disturbances. Fish must be allowed enough time to acclimate to the apparatus. Furthermore, metabolic rates in fish are sensitive to other environmental and behavioral factors including temperature, water chemistry, photoperiod, nutritional condition, and social interactions. Metabolic activity can also be determined by monitoring gill ventilation frequency, opercular muscle electrical activity, or heart rates.

Changes in growth rate are particularly important as indicators of chronic stress. In hatchery fish, effects on growth are usually calculated from after-the-fact changes in length or weight that include fat deposition as well as protein synthesis. Data on instantaneous growth rates due to effects on protein synthesis alone can be obtained by measured RNA:DNA ratios and uptake of radiolabeled amino acids in muscle or organ tissues.

Assessing effects of stress on reproduction is done by measuring changes in reproductive steroids, sperm index, oocyte atresia, and fecundity and hatching success (Donaldson 1990). Disturbances in the normal patterns of environmental factors, such as photoperiod, water temperature, or salinity, can influence gametogenesis and thus fecundity. In addition, some of the hormones involved in the stress response also participate in reproductive physiology. Thus, altered reproductive success may be due to chronic stress as well as to altered environmental parameters. Measuring blood concentrations of reproductive steroid hormones provides the most complete interpretation of effects of stress on reproduction (Pickering et al. 1987).

Selected physiological tests that can be used to evaluate the adverse effects of environmental stress and the tolerance limits for acclimation are given in Table 2.4. Clinical profiles of the blood and tissue chemistry changes that occur in

TABLE 2.4. Physiological tests and interpretive guidelines to assess effects of environmental factors on fish health and physiological condition (information compiled from Wedemeyer et al. 1990; Mayer et al. 1992).

Physiological Test	Diagnostic Significance if Results are: Low	High
Adenylate energy charge (muscle, liver)	Energy drains due to chronic stress	Normal bioenergetic conditions
Blood cell counts		
Erythrocytes	Anemias, hemodilution, impaired osmoregulation	Stress polycythemia
Leukocytes	Leukopenia due to acute stress	Leukocytosis due to bacterial infection
Thrombocytes	Abnormal blood clotting time	Thrombocytosis due to acute or chronic stress
Chloride (plasma)	Gill chloride cell damage, compromised osmoregulation	Hemoconcentration, compromised osmoregulation
Cholesterol (plasma)	Impaired lipid metabolism	Chronic stress, dietary lipidemia
Clotting time (blood)	Acute stress	No recognized significance
Cortisol (plasma)	No recognized significance	Acute or chronic stress
Gastric atrophy	No recognized significance	Chronic stress
Glucose (plasma)	Inanition	Acute or chronic stress
Glycine incorporation (scales)	Reduced growth	Normal conditions, good growth
Glycogen (liver, muscle)	Chronic stress, dietary problems	Liver damage, diet too high in carbohydrate
Hematocrit (blood)	Anemia, hemodilution	Hemoconcentration, gill damage
Hemoglobin (blood)	Anemias	Hemoconcentration, gill damage
Hemoglobin (mucus)	No recognized significance	Acute stress
Interrenal hypertrophy	No recognized significance	Chronic stress
Lactic acid (blood)	Normal conditions	Acute or chronic stress, muscular exertion
Osmolality (plasma)	External parasites, heavy metals, hemodilution	Dehydration, impaired osmoregulation
RNA/DNA (muscle)	Impaired growth, chronic stress	Normal conditions
Total protein (plasma)	Infection, nutritional problems, renal failure	Hemoconcentration

Note: control fish should always be tested for comparison.

response to stressful conditions can serve as leading indicators of more serious physiological dysfunctions such as impaired health, growth, or survival. Recommended physiological and performance capacity tests are discussed in Wedemeyer et al. (1990). Blood chemistry values typical of clinically healthy salmonid fishes are given in Table 2.5. When blood or tissue chemistry tests are used to diagnose fish health problems by any of the methods listed, control groups should always be included because of the limited information on normal values. A con-

TABLE 2.5. Representative blood and tissue chemistry values for clinically healthy fish under hatchery conditions. Rainbow trout, soft water, OMP diet; Atlantic salmon, salinity 32 ppt, commercial diet; channel catfish, soft water, commercial diet (compiled from information originally developed by Bentinck-Smith et al. 1987; Sandnes et al. 1988; Francis-Floyd 1992; and the research of the author).

Physiological Parameter	Channel Catfish	Rainbow Trout	Atlantic Salmon
Ascorbate, interrenal (μg/g)	—	102–214	—
Bicarbonate (meq/L)	—	8.9–15.9	—
Bilirubin (mg/dL)	1.6	0.4–4.5	—
Blood urea nitrogen (mg/dL)	2.6	0.9–4.5	—
CO_2 (volume %)	—	—	—
Calcium (mg/dL)	13.5	6.7–10.6	—
Chloride (meq/L)	131.5	84–132	—
Cholesterol (mg/dL)	151.7	161–365	9.3–12.8
Clotting time (s)	—	—	—
aorta canula	—	150–250	—
cardiac puncture	—	50–150	—
severed caudal peduncle	—	20–60	—
Cortisol (μg/dL)	—	1.5–18.5	—
Erythrocytes ($10^6/mm^3$)	2.4	0.6–1.3	0.85–1.1
Glucose (mg/dL)	64.5	41–151	—
Hematocrit (%)	40	24–43	44–49
Hemoglobin (g/dL)	4–8	5.4–9.3	8.9–10.4
Leukocytes ($1000/mm^3$)	28	5.4–36.0	—
Magnesium (meq/L)	4.2	1.2–3.3	—
Osmotic pressure (Mosm)	290	288–339	—
pH (10°C)	—	7.50–7.83	—
Phosphorous (mg/dL)	—	8.4–12.7	—
Thrombocytes ($1000/mm^3$)	—	11.0–21.0	—
Total protein (g/Dl)	2.2	2.0–6.0	4.2–5.6

siderable research effort still remains before clinical chemistry in fishes can be used routinely as it is in human and veterinary medicine. Values that appear normal may also simply mean that fish have regulated blood electrolytes, cortisol, glucose, hematological characters, and other variables back to normal or near-normal levels because acclimation has been achieved. The caloric cost of the stress response will still continue however.

Reliable data on the point at which the blood chemistry disturbances caused by stressful challenges become life threatening are not generally available for fish. The following information is intended only as a guideline. Hypochloremia is generally well tolerated by salmonids if blood chloride levels do not fall below about 90 meq/L. However, a decline in plasma osmotic pressure to below about 200 mOsm is diagnostic of compromised osmoregulation that may become life-threatening if not corrected. White blood cell counts (leukocytes) as low as 2,000/mm^3 apparently do not compromise disease resistance. Lower counts, however, may result in decreased ability to withstand infections. The commercially important salmonids used in aquaculture can survive surprising degrees of anemia. Blood hemoglobin concentrations down to about 2 g/dL and hematocrits down to about 15%, although not healthy, can be tolerated if the fish are not forced to swim actively. Lower values become life threatening.

REFERENCES

Adams, S. M., and J. E. Breck. 1990. Bioenergetics. Pages 389–409 *in* C. Schreck and P. Moyle (eds.), Methods for Fish Biology. American Fisheries Society, Bethesda, Maryland.

Amend, D. F., and L. Smith. 1974. Pathophysiology of infectious hematopoietic necrosis virus disease in rainbow trout (*Salmo gairdneri*): early changes in blood and aspects of the immune response after injection of IHN virus. Journal of the Fisheries Research Board of Canada 27:265–270.

Anderson, D. P. 1990. Immunological indicators: effects of environmental stress on immune protection and disease outbreaks. American Fisheries Society Symposium 8:38–50.

Anderson, D. P., and A. K. Siwicki. 1994. Duration of protection against *Aeromonas salmonicida* in brook trout immunostimmulated with glucan or chitosan by injection or immersion. Progressive-Fish Culturist 56:258–261.

Arkoosh, M. R., E. Casillas, E. Clemons, B. McCain, and U. Varanasi. 1991. Suppression of immunological memory in juvenile chinook salmon (*Oncorhynchus tshawytscha*) from an urban estuary. Fish & Shellfish Immunology 1:261–277.

Barton, B. A., and G. K. Iwama. 1991. Physiological changes in fish from stress in aquaculture with emphasis on the response and effects of corticosteroids. Annual Review of Fish Diseases 1:3–26.

Barton, B. A., and C. B. Schreck. 1987. Metabolic cost of acute physical stress in juvenile steelhead. Transactions of the American Fisheries Society 116:257–263.

Basu, S. P. 1959. Active respiration of fish in relation to ambient concentrations of oxygen and carbon dioxide. Journal of the Fisheries Research Board of Canada 16:175–212.

Beitinger, T. L., and R. W. McCauley. 1990. Whole animal physiological processes for the assessment of stress in fishes. Journal of Great Lakes Research 16:542–575.

Bentinck–Smith, J., M. H. Beleau, P. Waterstrat, C. S. Tucker, F. Stiles, P. R. Bowser, and L. A. Brown. 1987. Biochemical reference ranges for commercially reared channel catfish. The Progressive Fish-Culturist 49:108–114.

Blackburn, J., and W. C. Clarke. 1987. Revised procedure for the 24 hour seawater challenge test to measure seawater adaptability of juvenile salmonids. Canadian Technical Report in Fisheries and Aquatic Science 1515.

Bowser, P. R. 1993. Clinical pathology of salmonid fishes. Pages 327–332 *in* M. Stoskopf (ed.), Fish Medicine. W. B. Saunders, Philadelphia, Pennsylvania.

Bradley, T. B., and A. W. Rourke. 1984. The influences of addition of minerals to rearing water and smoltification on selected blood parameters of juvenile steelhead trout, *Salmo gairdneri* Richardson. Physiological Zoology 58:312–319.

Brett, J. R. 1973. Energy expenditure of sockeye salmon during sustained performance. Journal of the Fisheries Research Board of Canada 30:1799–1809.

Brown, C. R., and J. N. Cameron. 1991. The induction of specific dynamic action in channel catfish by infusion of essential amino acids. Physiological Zoology 64: 276–297.

Clarke, W. C. 1992. Environmental factors in the production of Pacific salmon smolts. World Aquaculture 23:40–42.

Colt, J. E., and G. Tchobanoglous. 1981. Design of aeration systems for aquaculture. Pages 138–148 *in* L. Allen and E. Kinney (eds.), Bioengineering Symposium for Fish Culture, American Fisheries Society, Bethesda, Maryland.

Condo, S. G., P. Arata, M. Corda, M. G. Pellegrini, A. Cau, and B. Giardina. 1985. International Symposium on Environmental Biogeochemistry; Viterbo (Italy), Rome, Italy.

Davison, W. 1989. Mini review: training and its effects on teleost fish. Comparative Biochemistry and Physiology 94A:1–10.

Dickhoff, W. C., C. L. Brown, C. V. Sullivan, and H. A. Bern. 1990. Fish and amphibian models for developmental endocrinology. The Journal of Experimental Zoology Supplement 4:90–97.

Donaldson, E. M., 1990. Reproductive indices as measures of the effects of environmental stressors in fish. American Fisheries Society Symposium 8:109–122.

Foote, C. J., I. Mayer, C. C. Wood, W. C. Clarke, and J. Blackburn. 1994. On the developmental pathway to nonanadromy in sockeye salmon, *Oncorhynchus nerka*. Canadian Journal of Zoology 72:397–405.

Francis-Floyd, R. 1992. Clinical pathology of catfishes. Pages 498–505 *in* M. Stoskopf (ed.), Fish Medicine, W. B. Saunders, Philadelphia, Pennsylvania.

Haswell, M. S., G. J. Thorpe, Harris, L. E., T. C. Mandis, and R. E. Rauch. 1982. Millimolar quantities of sodium salts used as prophylaxis during fish hauling. Progressive Fish-Culturist 44:179–182.

Heisler, N. 1993. Acid-base regulation. Pages 343–378 *in* D. H. Evans (ed.), The Physiology of Fishes. CRC Press, Boca Raton, Florida.

Hoar, W. S. 1988. The physiology of smolting salmonids. Pages 275–343 *in* W. Hoar and D. Randall (eds.), Fish Physiology, vol. XIB, Academic Press, San Diego.

Houston, A. H. and M. P. Schrap. 1994. Thermoacclimatory response: have we been using appropriate conditions and assessment methods? Canadian Journal of Zoology 72:1238–1242.

Iwama, G. K. 1992. Smolt quality. World Aquaculture 23:38–39.

Iwama, G. K., J. C. McGeer, N. J. Bernier. 1992. The effects of stock and rearing history on the stress response in juvenile coho salmon (*Oncorhynchus kisutch*). Pages 67–83 *in* C. Sindermann, B. Steinmetz, and W. Hershberger (eds.), Introduction and Transfer of Aquatic Species—Selected Papers. International Atlantic salmon Foundation, Halifax, Nova Scotia.

Liao, P. 1971. Water requirements of salmonids. Progressive Fish-Culturist 38:210–218.

Maule, A. G., C. B. Schreck, and S. L. Kaattari. 1987. Changes in the immune system of coho salmon (*Oncorhynchus kisutch*) during the parr-to-smolt transformation and after implantation of cortisol. Canadian Journal of Fisheries and Aquatic Sciences 44:161–166.

Maule, A. G., C. B. Schreck, C. S. Bradford, and B. A. Barton. 1988. Physiological effects of collecting and transporting emigrating juvenile chinook salmon past dams on the Columbia River. Transactions of the American Fisheries Society 117:245–261.

Maule, A. G., R. A. Tripp, S. L. Kaattari, and C. B. Schreck. 1989. Stress-induced changes in glucocorticoids alter immune function and disease resistance in chinook salmon (*Oncorhynchus tshawytscha*). Journal of Endocrinology 120:135–142.

Mayer, F. L., D. J. Versteeg, M. J. McKee, L. C. Folmar, R. L. Graney, D. C. McCume, and B. A. Rattner. 1992. Physiological and nonspecific biomarkers. Pages 5–83 *in* R. J. Huggett (ed.), Biomarkers: Biochemical, Physiological, and Histological Markers of Anthropogenic Stress. Lewis Publishers, Boca Raton, Florida.

McDonald, D. G., M. D. Goldstein, and C. Mitton. 1993. Response of hatchery-reared brook trout, lake trout, and splake to transport stress. Transactions of the American Fisheries Society 122:1127–1138.

Peters, G., H. Delventhal, and H. Klinger. 1981. Stress diagnosis for fish in intensive culture systems. Pages 239–248 *in* K. Tiews (ed.), Aquaculture in Heated Effluents and Recirculation Systems, vol. 2. Heenemann Verlagsgesellschaft, Berlin.

Pickering, A. D., T. G. Pottinger, J. Carragher, and J. P. Sumpter. 1987. The effects of acute and chronic stress on the levels of reproductive hormones in the plasma of mature male brown trout, *Salmo trutta* L. General and Comparative Endocrinology 68:249–259.

Randall, D. J., and C. Daxboeck. 1984. Oxygen and carbon dioxide transfer across fish gills. Pages 263–314 *in* W. Hoar and D. Randall (eds.), Fish Physiology, vol. X. Academic Press, San Diego.

Rice, J. A. 1990. Bioenergetics modeling approaches to evaluation of stress in fishes. American Fisheries Society Symposium 8:80–92.

Rousset, V., and A. Raibaut. 1984. Anatomical and functional effects of *Pharodes banyulensis* infections on *Blennius pavo* (Pisces, Teleostei, Blenniidae) in a French mediterranean pond (Bassin de Thau). Zeitschrift Parasitenkd. 70:119–130.

Sandnes, K., Lie, O. L., and R. Waagbo. 1988. Normal ranges of some blood chemistry parameters in adult farmed Atlantic salmon, *Salmosalar.* Journal of Fish Biology 32:129–136.

Saunders, R. L., and J. Dustin. 1992. Increasing production of Atlantic salmon smolts by manipulating photoperiod and temperature. World Aquaculture 23: 43–46.

Saunders, R. L., E. B. Henderson, P. R. Harmon, C. E. Johnson, and J. G. Eales. 1983. Effects of low environmental pH on smolting of Atlantic salmon (*Salmo salar*). Canadian Journal of Fisheries and Aquatic Sciences 40:1203–1211.

Schreck, C. B. 1981. Stress and compensation in teleostean fishes: response to social and physical factors. Pages 295–321 *in* A. Pickering (ed.), Stress and Fish. Academic Press, London.

Schreck, C. B., A. G. Maule, and S. L. Kaattari. 1989. Stress and disease resistance. Pages 177–181 *in* M. Carrillo, S. Zanuy, and J. Planas (eds.), Proceedings of Satellite Symposium on Applications of Comparative Endocrinology to Fish Culture. Imprente Vilaro, Barcelona, Spain.

Shrimpton, J. M., and D. J. Randall. 1992. Smolting and survival in wild versus hatchery coho salmon. World Aquaculture 23: 51–54.

Slater, C. H., and C. B. Schreck. 1993. Testosterone alters immune response of chinook salmon (*Oncorhynchus tshawytscha*). General and Comparative Endocrinology 89:291–298.

Smith, L. S. 1982. Introduction to Fish Physiology. TFH Publications, Neptune, New Jersey.

Stoskopf, M. K. 1993. Fish Medicine. W. B. Saunders, Philadelphia, Pennsylvania.

Thomas, P. 1990. Molecular and biochemical responses of fish to stressors and their potential use in environmental monitoring. American Fisheries Society Symposium 8:9–28.

Thomas, S., and S. F. Perry. 1992. Control and consequences of adrenergic activation of red blood cell Na^+/H^+ exchange on blood oxygen and carbon dioxide transport in fish. Journal of Experimental Zoology 263:160–175.

Thorarensen, H., P. E. Gallaugher, A. K. Kiessling, and A. P. Farrell. 1993. Intestinal blood flow in swimming chinook salmon *Oncorhynchus tshawytscha* and the effects of haematocrit on blood flow distribution. Journal of Experimental Biology 179: 115–129.

Tucker, C. S., and E. H. Robinson. 1990. Channel Catfish Farming Handbook. Van Nostrand Reinhold, New York.

Urawa, S. 1993. Effects of *Ichthyobodo necator* infections on saltwater survival of juvenile chum salmon (*Oncorhynchus keta*). Aquaculture 110:101–110.

Wedemeyer, G. A., and C. P. Goodyear. 1984. Diseases caused by environmental stressors. Pages 424–434 *in* O. Kinne (ed.), Diseases of Marine Animals, vol. 4, part 1: Introduction, Pisces. Biologische Anstalt Helgoland, Hamburg, FRG.

Wedemeyer, G., and W. T. Yasutake. 1977. Clinical methods for the assessment of the effects of environmental stress on fish health. Technical Paper 89, U.S. Department of the Interior, Fish and Wildlife Service, Washington, D.C.

Wedemeyer, G. A., F. P. Meyer, and L. Smith. 1976. Environmental Stress and Fish Diseases. TFH Publications, Neptune, New Jersey.

Wedemeyer, G. A., R. L. Saunders, and W. C. Clarke. 1980. Environmental factors affecting smoltification and early marine survival of anadromous salmonids. U.S. National Marine Fisheries Service Marine Fisheries Review 42(6):1–14.

Wedemeyer, G. A., B. A. Barton, and D. J. McLeay. 1990. Stress and acclimation. Pages 451–489 *in* C. Schreck and P. Moyle (eds.), Methods for Fish Biology. American Fisheries Society, Bethesda, Maryland.

Weeks, B. A., J. E. Warinner, and C. D. Rice. 1989. Recent advances in the assessment of environmentally–induced immunomodulation. Oceans 89:408–411.

Wendt, C. A. G., and R. L. Saunders. 1973. Changes in carbohydrate metabolism in young Atlantic salmon in response to various forms of stress. Pages 55–82 *in* M. Smith and W. Carter (eds.), Proceedings of the International Symposium on the Atlantic Salmon: Management, Biology, and Survival of the Species. Unipress, Fredericton, N. B., Canada.

Wilson, R. P., and J. E. Halver. 1986. Protein and amino acid requirements of fishes. Annual Reviews of Nutrition 6:225–244.

Yasutake, W. T., and J. H. Wales. 1983. Microscopic Anatomy of Salmonids: An Atlas. U.S. Fish and Wildlife Service Resource Publication 150.

Young, P. S., and J. J. Cech Jr. 1993. Improved growth, swimming performance, and muscular development in exercise-conditioned young-of-the-year striped bass (*Morone saxatilis*). Canadian Journal of Fisheries and Aquatic Sciences 50:703–707.

3
Interactions with Water Quality Conditions

INTRODUCTION

Water quality is widely acknowledged to be one of the most important rearing conditions that can be managed to reduce disease exposure and stress in intensive fish culture. However, the physiological tolerance of fish to water quality alterations is affected by a number of environmental and biological variables and it is not a simple matter to identify specific chemical constituents, temperatures, or dissolved gas concentrations that will provide optimum rearing conditions under all circumstances. First, the effects of water quality conditions on fish health vary considerably with species, size, age, and previous history of exposure to each chemical constituent in question. Second, the water quality conditions themselves (particularly pH, dissolved oxygen, and temperature) can greatly alter the biological effects of dissolved substances. For example, concentrations of zinc and copper that can cause lethal gill damage in soft, acidic water become nearly nontoxic in hard, alkaline water (pH >7, total hardness >200 mg/L as $CaCO_3$). Similarly, water temperature and DO affect heavy metal toxicity by acting to increase or decrease gill ventilation rates and hence the total amount of toxicant the gill epithelium is physically exposed to. Finally, most water quality information has been developed to define the acute and chronic toxicity levels, not to define the concentrations that will provide optimum rearing conditions. Thus, there is ample information available on the concentrations of heavy metals that are acutely lethal (e.g., the 96-h LC_{50}); somewhat less information on maximum safe chronic exposure levels that will cause no deleterious effects over a life cycle; and very little data on the dissolved metal concentrations required to promote physiological health and disease resistance. For example, two of the toxic heavy metals (zinc, copper) are actually required dietary micronutrients for normal growth and development. Two other potential toxicants, fluoride and selenium, may be involved in resistance to bacterial kidney disease (Lall et al. 1985). However, there is little information on what the optimum dissolved concentrations of any of these ions might be.

Despite the complex issues involved, widespread agreement exists on some of the water quality conditions necessary to support the health of fish in intensive culture. The following guidelines would probably be considered desirable for both cold- and warmwater species:

1. dissolved oxygen, 7 mg/L minimum
2. pH, 7–8 average; extremes, 6–9
3. un-ionized ammonia (NH_3), 0.05 ppm maximum
4. total CO_2 <10–15 mg/L.

In practice, effective fish health management requires species specific and more detailed water quality information, including maximum safe exposure levels for the contaminants and other toxic substances that can sometimes occur in hatchery water supplies. A summary of such water quality requirements for (anadromous) salmonids is presented in Table 3.1. These data provide the basic technical framework upon which disease prevention through environmental management must be based. In the following sections, the most important of the individual water quality factors listed in Table 3.1 will be discussed together with a consideration of specific fish health problems associated with natural chemical constituents, contaminants, and algal toxins.

WATER QUALITY REQUIREMENTS

Acidity

Acidity is the capacity of water to neutralize hydroxyl ions (OH^-). Strictly speaking, it should be measured by titration with a standard base like 0.02 N NaOH to the phenolphthalein end point (pH 8.2) and expressed as meq/L. In aquacultural work, however, it is more convenient to express acidity in terms of its intensity, rather than as a capacity factor, and so the pH is often measured instead. A pH of <7.0 is acidic, pH 7 is neutral, pH >7 is alkaline.

Acidity in waters used for fish culture is generally due to carbon dioxide (CO_2) dissolved from the atmosphere or produced by fish metabolism, mineral acids from pollution (e.g., acid precipitation, acid mine drainage[5]), naturally occurring

5. From coal or other mineral mining operations.

TABLE 3.1. Water chemistry limits recommended to protect the health of cold- and warmwater fish in intensive culture (data compiled from Wedemeyer and Goodyear 1984; Post 1987; Klontz 1993; and the experience of the author).

Parameter	Recommended Limits
Acidity	pH 6–9
Arsenic	<400 µg/L
Alkalinity	>20 mg/L (as $CaCO_3$)
Aluminum	<0.075 mg/L
Ammonia (un-ionized)	<0.02 mg/L
Cadmium[a]	<0.0005 mg/L in soft water; <0.005 mg/L in hard water
Calcium	>5 mg/L
Carbon dioxide	<5–10 mg/L
Chloride[b]	>4.0 mg/L
Chlorine	<0.003 mg/L
Copper[a]	<0.0006 mg/L in soft water <0.03 mg/L in hard water
Gas supersaturation	<110% total gas pressure (103% salmonid eggs/fry; 102% lake trout)
Hydrogen sulfide	<0.003 mg/L
Iron	<0.1 mg/L
Lead	<0.02 mg/L
Mercury	<0.0002 mg/L
Nitrate (NO_3^-)	<1.0 mg/L
Nitrite (NO_2^-)	<0.1 mg/L
Oxygen	6 mg/L, coldwater fish 4 m/L, warmwater fish
Selenium	<0.01 mg/L
Total Dissolved Solids	<200 mg/L
Total Suspended Solids	<80 mg/L
Turbidity (NTU)	<20 NTU over ambient levels
Zinc	<0.005 mg/L

[a] To protect smolt development of anadromous salmonids: hard water >100 mg/L total hardness (as CaCO3), soft water <100 mg/L.

[b] To protect against nitrite toxicity in reuse systems using biofilters for ammonia removal.

organic acids from humus deposits, or the hydrolysis of salts leached into water supplies from mineral deposits. Mine drainage water becomes acidic because iron bacteria oxidize pyrites (FeS_2) and other sulfide impurities in the ore to sulfuric acid (H_2SO_4). In hatcheries supplied with hard, alkaline water (>200 mg/L as $CaCO_3$), introduced acidic compounds such as acid mine waste drainage may be partly or completely neutralized by the natural carbonate content of the water. However, dissolved CO_2 is then produced that can be as deleterious to fish health as the low pH of the acid mine waste itself. Carbonate rock can also be placed in the mine waste drainage stream to neutralize H_2SO_4 but this may result in the production of semigelatinous hydrated iron oxide ($Fe(OH)_3$) that may physically damage the gills.

The CO_2 produced by fish respiration itself can substantially lower the pH in hatcheries using water low in total hardness (<50 mg/L as $CaCO_3$). This condition is rarely a problem in raceways but can be a problem in fish transport operations where CO_2 concentrations of 20–30 mg/L can accumulate in the tanks, lowering the pH to 6.0 or less within 30 min of loading (discussed in Chapter 4). The pH can fluctuate by one or two units even in well-buffered ponds (alkalinity >100 mg/L) because of the CO_2 produced by fish respiration.

Acidity in hatchery water supplies has several adverse effects on fish health. If severe, ion transport at the gills is affected, leading to osmoregulatory failure and death. Increased susceptibility to infectious diseases may also occur but there is insufficient data from controlled studies to provide guidelines for corrective action. A better documented deleterious effect of acidity is the solubilization of toxic trace metals in soils by acid rain. This solubilization allows previously bound metals to leach into the water column. Aluminum toxicity has become an important present day example of this problem. At pH 6.6, 30-day aluminum exposures are safe for the egg and fry stages of brook trout at concentrations up to 57 μg/L (ppb). At pH 5.6, this concentration decreases to only 29 ppb (Cleveland et al. 1989). Acidity also interferes with the normal development of smoltification in juvenile anadromous salmonids, even in the absence of aluminum toxicity. The parr–smolt transformation of juvenile Atlantic salmon is normal at pH 6.4–6.7, but seriously impaired at pH 4.2–4.7 (Saunders et al. 1983). This can happen when acid precipitation accumulates in the snow pack causing spikes of acidity in freshwater streams during the spring runoff during the critical parr–smolt transformation of resident anadromous salmonids.

Upper and lower lethal limits for pH are not fixed values but vary somewhat depending on other environmental factors such as temperature, aluminum and calcium concentrations, and acclimation pH. Fish populations in natural waters can tolerate the pH extremes of 5–9 (absent the effects of aluminum toxicity) but a more prudent guideline to protect the health of freshwater fish in intensive culture is pH 6.5–9.0. Fish in mariculture facilities such as net pens normally do not experience acidity because seawater is strongly buffered at a pH of about 8.2.

Alkalinity

Alkalinity is a measure of the total concentration of the alkaline (basic) substances dissolved in water. Like acidity, it is a capacity factor—the capacity of water to neutralize acids (hydrogen ions H^+). Alkalinity is measured by titration with standardized acid to the methyl orange end point (pH 4.3) and expressed as meg/L (as $CaCO_3$). Most waters of high alkalinity also have an alkaline pH (pH >7) and a high concentration of total dissolved solids (TDS). In fisheries work, alkalinity is sometimes (incorrectly) expressed in terms of pH as well as in mg $CaCO_3$/L.

Alkalinity in hatchery water supplies is due for the most part to naturally occurring dissolved bicarbonates (HCO_3^-), carbonates (CO_3^{-2}), and mineral hydroxides, although industrial and municipal effluents and agricultural drain water can also contribute.

The alkalinity of a hatchery water supply has direct and indirect effects on fish health. Alkalinity provides the buffering capacity needed to protect intensively cultured fish against the wide swings in water pH that otherwise would occur because of their own respiration and that of aquatic plants. In the absence of pollution, water with high alkalinity also has a high concentration of inorganic carbon and therefore serves as a measure of biological productivity, that is, the ability of water to support aquatic plant and thus aquatic animal life. In warmwater pond-fish culture, the water is often soft and poorly buffered. Rapid algal growth can consume the carbon dioxide faster than it can be replaced by equilibration from the atmosphere. Oxygen is released into the water but the pH can increase to 9.5 or even 10 in a matter of hours. These pH increases are temporary and usually well tolerated by warmwater fishes. However, salmonid fishes exposed to a pH of 9–10 in high alkalinity waters can suffer inhibited ammonia/sodium exchange at the gill resulting in mortalities from the resulting high blood ammonia levels (Wright and Wood 1985). If ammonia is present, the high pH will greatly increase its toxicity to both cold- and warmwater species (discussed next). Sodium bicarbonate at 10–20 lbs per acre is sometimes added to warmwater fish ponds to temporarily correct low alkalinity and correct CO_2 and NH_3 problems arising from low or high pH.

For intensive fish culture, an alkalinity of 100–150 mg/L is recommended to provide the buffering capacity needed to prevent wide pH fluctuations (pH extreme of 9.0), promote algae production, prevent leaching of heavy metals, and to allow the use of copper compounds for disease treatments.

Ammonia

Trace concentrations of ammonia (NH_3) can often be found in hatchery water supplies due to the decay of nitrogenous organic material in soils or sediments. Larger amounts of ammonia in ground water or surface water usually indicate pollution from sewage treatment plants, drainage from agricultural lands, or industrial effluent discharges. In most ponds or raceways however, the major source of ammonia is fish metabolism. Relatively large amounts of ammonia (and carbon dioxide) together with smaller amounts of urea, creatine, creatinine, uric acid, and other nitrogenous wastes are more-or-less continuously excreted. For example, channel catfish excrete about 20 g of ammonia per day per kg of ration fed. Salmonids fed typical dry diets excrete 25–35 grams of ammonia per kg of feed consumed (Colt and Armstrong 1981). In raceway rearing units, the water exchange rate is adjusted to carry the ammonia away. In water reuse systems with biological filtration, nitrifying bacteria are used to remove the ammonia by oxidation to nitrite (NO_2^-) and then to nitrate (NO_3^-). In properly operating ponds, most of the ammonia is assimilated and removed by algae and aquatic plants. Ammonia can also be removed from fish culture systems by ion exchange using clinoptilolite, a naturally occurring zeolite mineral. The absorptive capacity can be as high as 9 mg ammonia removed per gram of clinoptilolite (Marking and Bills 1982).

When ammonia enters the water, the hydrogen ions present immediately react and convert it into an equilibrium mixture of the essentially nontoxic ammonium ion (NH_4^+), and un-ionized NH_3, which is toxic.

$$NH_3 + H^+ + OH^- \rightleftarrows NH_4^+ + OH^-$$

The amount of toxic NH_3 present is of most interest to fish culturists. However, chemical analysis of the water measures the sum of both forms of ammonia present (i.e., NH_3 and NH_4^+) and the actual amount of toxic NH_3 present must be calculated. Its concentration depends primarily on the pH and, to a lesser extent, the water temperature, salinity, and total dissolved solids concentration. Based on the water pH, the amount of NH_3 present is easily calculated from the equilibrium equation

$$\%\ NH_3 \rightleftarrows 100/(1 + \text{antilog}(pK_a - pH)).$$

For convenience, NH_3 values calculated from this equilibrium equation have been tabulated in Table 3.2 over the range of temperature and pH of interest in fish culture. The pK_a for the reaction is not shown but is relatively high (about 9.7 at 10°C). Thus, little of the total ammonia present will be in the toxic NH_3 form unless the water is quite alkaline (pH $\ll$7). In the soft, acidic water typical of salmonid aquaculture (pH $<$7), less than 1% of any ammonia present will be in its toxic un-ionized form. Thus, fish culturists working with hard alkaline water supplies generally face the most serious ammonia toxicity problems. As a practical example, consider that the total ammonia concentration in a fish hauling tank can easily reach 10 mg/L within an hour of loading. As seen from Table 3.2, if the pH is 8.0 and the water temperature is 10°C, the concentration of toxic NH_3 in the hauling tank will be about 0.18 mg/L; well above the 4–hour maximum safe exposure

TABLE 3.2. Percent of total ammonia that will be in the toxic NH3 form over the range of pH and water temperature of importance in freshwater fish culture operations.

	Water temperature (°C)			
pH	5	10	15	20
6.0	0.01	0.02	0.03	0.04
6.2	0.02	0.03	0.04	0.06
6.4	0.03	0.05	0.07	0.10
6.6	0.05	0.07	0.11	0.16
6.8	0.08	0.12	0.17	0.25
7.0	0.13	0.18	0.27	0.40
7.2	0.20	0.29	0.43	0.63
7.4	0.32	0.47	0.69	1.0
7.6	0.50	0.74	1.08	1.60
7.8	0.79	1.16	1.71	2.45
8.0	1.24	1.83	2.68	3.83
8.2	1.96	2.87	4.18	5.93
8.4	3.07	4.47	6.47	9.09
8.6	4.78	6.90	9.88	13.68
8.8	7.36	10.51	14.80	20.07
9.0	11.18	15.70	21.59	28.47

Source: Calculated from % $NH_3 = 100/(1 + \text{antilog}(pK_a - pH))$ using pK_a values from standard reference works.

Note: If the total dissolved solids (TDS) is >400–500 mg/L, NH_3 concentrations will be 10–15% lower (Messer et al. 1984).

level of 0.1 mg/L presently recommended for salmonids (EPA 1986). However, if the pH is 7 (at the same water temperature), the concentration of NH_3 present will only be 0.02 mg/L—well within safe limits.

In applying these calculations, the correction for TDS given in Messer et al. (1984) may also need to be applied. The amount of NH_3 formed will be about 10–15% lower in hard water areas. For seawater aquaculture operations, a separate correction for salinity is required (Table 3.3).

Other water quality factors may also act to change the toxicity of un-ionized ammonia after it is formed. Aside from DO, the effects of the dissolved sodium (Na^+) concentration and the total hardness are especially noteworthy. Because of the gill NH_4^+/Na^+ ion exchange mechanism (discussed in Chapter 2), abnormally low dissolved sodium concentrations in the hatchery water supply can result in fish with chronically elevated blood ammonia levels. Salt (NaCl) additions to increase the Na^+ concentration to 10–20 mg/L have been used to reduce chronically elevated blood ammonia levels in one soft water salmon hatchery (Bradley and Rourke 1985). Hard alkaline waters (pH 9–10) can also inhibit gill NH_4^+/Na^+ ex-

TABLE 3.3. Percent of total ammonia that will be in the unionized (toxic) form in seawater of salinity 28–31 ppt at pH and water temperatures of importance to aquaculture.

	Water temperature (°C)			
pH	5	10	15	20
7.2	0.17	0.24	0.35	0.51
7.4	0.26	0.38	0.56	0.81
7.6	0.42	0.60	0.88	1.27
7.8	0.66	0.95	1.39	2.00
8.0	1.04	1.49	2.19	3.13
8.2	1.63	2.34	3.43	4.88
8.4	2.56	3.66	5.32	7.52
8.6	4.00	5.68	8.18	11.41
8.8	6.20	8.72	12.38	16.96
9.0	9.48	13.15	18.29	24.45

Source: Calculated from $\%NH_3 = 100/(1 + \text{antilog}(pK_a - pH))$ using pK_a values from standard reference works.

change and elevate blood ammonia concentrations to toxic levels (Wright and Wood 1985).

The maximum safe exposure level for un-ionized ammonia has not been completely defined for fish under intensive culture conditions but guidelines do exist (Meade 1985). For anadromous salmonids in raceway culture, a concentration of 0.02–0.03 mg NH_3/L is probably about the maximum that should be allowed to prevent health problems such as environmental gill disease. Many fish culturists prefer to keep concentrations at or below 0.01 mg/L.[6] For acute exposures, such as in fish hauling operations, concentrations up to 0.1 mg/L are usually well tolerated by both warm- and coldwater species.

Carbon Dioxide

Most surface waters contain a small amount (1–2 mg/L) of carbon dioxide (CO_2) that has been dissolved from the atmosphere, produced by the microbial decomposition of organic matter in bottom sediments, or by the respiration of microorganisms, algae, and other aquatic plants. CO_2 levels can be higher when water from springs or wells is used, or when acid precipitation or mine drainage has reacted with dissolved carbonate minerals. Usually, however, the major source

6. If ammonia is to be expressed as ammonia-nitrogen instead of ammonia, the conversion factor $NH_3 = 1.21 \times NH_3\text{–}N$ should be used.

of CO_2 in ponds and raceways is fish metabolism. Salmonids, for example, expire about 1.4 mg of CO_2 for each mg of oxygen that they consume (Colt and Tchobanoglous 1981). Thus, the total CO_2 production by the fish in a rearing unit can be quite large even at only average swimming activity and oxygen consumption levels (see Table 2.1, Chapter 2) and carbon dioxide will accumulate rather quickly unless steps are taken to prevent it. In ponds and raceways, carbon dioxide is continuously removed by aquatic plants and/or the water exchange rate. In fish hauling tanks, concentrations of 20–30 mg/L are not unusual within 30 minutes after loading.

In contrast to the other atmospheric gases important to aquaculture (O_2 and N_2), carbon dioxide reacts with water when it dissolves forming a mixture of CO_2, carbonic acid (H_2CO_3), and nontoxic bicarbonate (HCO_3^-) and carbonate (CO_3^{-2}) ions.

$$CO_2 + H_2O \rightleftarrows H_2CO_3 \rightleftarrows H^+ + HCO_3^- \rightleftarrows H^+ + CO_3^{-2}$$

The relative proportions of each species present and thus the toxicity of dissolved CO_2 is determined primarily by the pH. Below pH 5, most of the dissolved carbon dioxide exists as CO_2; between pH 7–9, as nontoxic HCO_3^-; and above pH 11, as the CO_3^{-2} ion. The amount of H_2CO_3 present at any one time is small and is usually neglected. In alkaline water, toxicity is reduced because the dissolved CO_2 is partially converted to the nontoxic bicarbonate and carbonate ions. In acidic water, more toxic CO_2 is present but it is also fairly easily removed from solution by moderate mechanical agitation or aeration provided that surface agitation also occurs.

The CO_2 concentration can be measured by direct titration or calculated from the pH, alkalinity, and water temperature because these three variables are interrelated. Nomographs for the numerical factors required can be found in standard reference works (Tucker 1993). Regardless of method used, chemical analysis measures the total amount of carbon dioxide present. The relative concentrations of CO_2, bicarbonate, carbonate, and carbonic acid must be calculated separately.

The adverse effects of carbon dioxide on fish health are affected by environmental conditions such as the DO and water temperature. Low DO increases CO_2 toxicity whereas increased water temperature decreases it by decreasing its solubility. For coldwater fishes such as salmonids, the detrimental effects of CO_2 begin at carbon dioxide concentrations of >20 mg/L. As ambient CO_2 levels rise, blood CO_2 increases (hypercapnia) and the oxygen carrying capacity of hemoglobin begins to decrease due to the Bohr effect. Respiratory distress from reduced oxygen transport to the tissues occurs at ambient CO_2 levels >40 mg/L. Severe acidosis culminating in CO_2 narcosis (anesthesia) and death occurs as CO_2 levels exceed about 100 ppm[7]. The magnitude and relationship of these effects in salmonids were first

---

7. Severe acidosis is probably more often due to elevated blood lactic acid concentrations (hyperlacticemia) because of excitement and excessive swimming activity than to hypercapnia.

documented by Basu (1959) who showed that the DO required to provide sufficient oxygen to the tissues of salmonids to support a moderate swimming activity level increased from about 6 mg/L if little or no CO_2 was present, to more than 11 mg/L if the dissolved CO_2 concentration was 30 mg/L. Thus, the usual recommendation that fish in intensive culture will have adequate oxygen so long as the DO does not fall below about 80% of saturation must be qualified; at least for coldwater species. If dissolved CO_2 levels are not kept well below 30–40 mg/L, the blood oxygen carrying capacity can become depressed to the point that even vigorous aeration may be insufficient to prevent a degree of tissue hypoxia. As mentioned, aeration systems can help remove CO_2 but must be designed to provide sufficient surface agitation and headspace ventilation or the CO_2 will simply redissolve. Because CO_2 is highly soluble, only a small increase in CO_2 in the headspace gas will dramatically reduce CO_2 stripping efficiency (Watten et al. 1991).

Experience has shown that fish under hatchery conditions can tolerate chronically elevated dissolved CO_2 concentrations in the range of 15–20 mg/L, if the rate of increase is slow. Under these circumstances the blood buffering system can stabilize the blood pH and prevent acidosis by increasing the plasma bicarbonate concentration sufficiently to compensate for the increased blood CO_2. If the hypercapnia occurs too rapidly for physiological compensation to occur, as in fish transport operations, millimolar quantities of sodium bicarbonate or sodium sulfate can be added to the hauling tank water. This will partially offset the decrease in blood buffering capacity caused by the hypercapnia and help maintain the blood pH (Haswell et al. 1982).

To ensure good health and physiological condition, warmwater species should not be exposed to dissolved CO_2 concentrations higher than 20–30 mg/L for extended periods. For salmonid fishes, CO_2 concentrations <10–20 mg/L are desirable. During fish hauling operations, CO_2 concentrations of up to 30–40 mg/L are well tolerated by both warm- and coldwater species, if the DO is at saturation, because of the short time periods involved.

Chlorine

Chlorine is not a natural constituent of surface or groundwater and therefore should never be found in hatchery water supplies unless they have been contaminated by sources such as sewage treatment plants or power plant cooling systems. However, chlorine does play an important role in intensive fish culture. Chlorine is routinely used to disinfect tanks and equipment, water discharged from fish facilities may be chlorinated to destroy pathogens, and municipal water is sometimes dechlorinated so it can be used for fish rearing.

Chlorine for disinfection is usually applied as a sodium or calcium hypochlorite solution ($NaOCl$, $Ca(OCl)_2$), as granular calcium hypochlorite (HTH), or as chlorine gas (Cl_2). For treating large volumes of water, gas chlorination is the method of choice because of its lower cost.

When sodium or calcium hypochlorite are added to water, they react to produce a mixture of strongly germicidal and toxic hypochlorous acid (HOCl) together with sodium or calcium hydroxide. Disinfectant solutions made with hypochlorite salts tend to be alkaline because the hydroxides dissociate liberating (OH^-) ions.

$$NaOCl + H_2O \rightleftarrows NaOH + HOCl$$
$$Ca(OCl)_2 + H_2O \rightleftarrows Ca(OH)_2 + HOCl$$
$$Ca(OH)_2 \rightleftarrows Ca^{+2} + OH^-$$

Chlorine gas added to water reacts to form a mixture of HOCl, hydrogen ions (H^+), and nontoxic chloride ions (Cl^-).

$$Cl_2 + H_2O \rightleftarrows H^+ + Cl^- + HOCl$$
$$HCl \rightleftarrows H^+ + Cl^-$$

Gas chlorination tends to acidify water because of the hydrogen ions produced.

Regardless of the chemical form of chlorine applied, the toxic HOCl produced immediately dissociates forming an equilibrium mixture of H^+, HOCl, and less toxic OCl^-; the relative proportions of which are determined by the pH (discussed further in Chapter 6).

$$HOCl \rightleftarrows H^+ + OCl^-$$

In the pH 6–7 range, typical of most soft water used for rearing anadromous salmonids, the highly toxic HOCl is the major species present. In alkaline water, more of the less toxic OCl^- will be present. Above pH 9, almost all of the chlorine will be in the less toxic OCl^- form. The individual concentrations of OCl^- and HOCl are somewhat difficult to measure and methods used to analyze for chlorine measure their sum, usually termed the *free residual (or total) chlorine concentration*.

In addition to reacting with water to form HOCl and OCl^-, chlorine will also react with any ammonia or other nitrogenous materials present in the water forming chloramine compounds. Paradoxically, chloramines are weaker germicides than chlorine, but their fish toxicity is usually greater. In chlorine determinations, the concentration of chloramines themselves is termed the *combined chlorine residual*. Again, separating individual concentrations is somewhat difficult and commonly used analytical methods, such as the *o*-toluidine determination, measure the total of the free chlorine and the combined chlorine concentrations. This measurement will overstate the germicidal activity and understate the fish toxicity if significant amounts of chloramines are present.

In the absence of chloramines, the fish toxicity of chlorine is due primarily to its HOCl and OCl^- forms. Both are strong oxidants and destroy gill tissue by penetrating cell membranes and causing nonspecific damage to cell structures, enzymes, and the purine and pyrimidine ring structure of DNA and RNA. The OCl^- ion has about the same molecular weight as HOCl but its electrical charge impedes penetration through cell membranes making it slightly less toxic to fish.

The ability of chlorine to disinfect is also due to HOCl and OCl^-. Again, both are strong oxidants but HOCl is a stronger germicide because the electrical charge of the OCl^- ion impedes its penetration into microbial cells. Thus, the pH has a very important effect on disinfection because it controls the amount of HOCl present. Above pH 9, at least 96% of the HOCl is dissociated into OCl^-, which is only weakly germicidal whereas at pH 6, about 96% of the chlorine will be in the more effective HOCl form. In hard water areas, sufficient (acetic) acid to reduce the pH to about 6 should be added to disinfectant solutions to maximize the formation of HOCl. Suspended particulates such as clay or organic matter tend to protect microorganisms from direct chlorine exposure but dissolved minerals generally have little effect.

Chlorination to destroy pathogens in effluents from fish holding facilities is discussed in Chapter 6, but briefly a free residual chlorine concentration of 1–3 mg/L with a contact time of 10–15 minutes is recommended. If exotic pathogens are involved, residuals of 3–5 mg/L are suggested with a contact time of 15–30 minutes to allow a margin of safety. For equipment and tank disinfection, a chlorine concentration of 100–200 mg/L with a contact time of at least 30 minutes is recommended. For convenience, the chlorine is usually applied as a NaOCl solution or as granular $Ca(OCl)_2$. For fish transport tanks, sufficient water to cover the intakes of the recirculating pumps or spray agitator propellers is poured in and the calculated amount of liquid or dry chlorine added (e.g., 15 g HTH, 70% available chlorine, per 100 L). If the water is alkaline, sufficient acetic acid to decrease the pH to about 6 should be added to activate the chlorine. The solution is pumped through the system for at least 30 min with air or oxygen flowing through the diffusers. After disinfection, the equipment must be thoroughly rinsed with chlorine-free water and any residual chlorine allowed to dissipate naturally (24–48 hours), or neutralized with a sodium thiosulfate rinse, before it is used to handle fish. The concentration of sodium thiosulfate needed is 7.4 ppm per ppm chlorine to be neutralized (Jensen 1989). As a precaution, commercially available chlorine test paper or a chemical method such as the *o*-tolidine determination should always be used to confirm that chlorine residuals have dissipated to safe levels or have been adequately neutralized.

Chlorine can be neutralized by treatment with sulfur compounds (sulfur dioxide gas, sodium thiosulfate, sodium sulfite), removed by filtration through activated carbon, or allowed to dissipate naturally. The latter is too variable to be completely safe. As a guideline about 20 hours per mg/L of chlorine to be removed should be allowed.

For large volumes of water, neutralization with sulfur dioxide (SO_2) gas is the most practical. SO_2 reacts with water to form sulfite ions (SO_3^{-2}) which react with both chlorine and chloramines forming nontoxic chloride ions, ammonium salts, and small amounts of acid in a 1:1 stoichiometric ratio. For smaller scale applications, a sodium sulfite (Na_2SO_3) or thiosulfate ($Na_2S_20_3$) solution is usually added. These compounds hydrolyze to produce sulfite ions which then react with chlorine in the same way as SO_2. As mentioned, a concentration of about 7.4 ppm of sodium thiosulfate pentahydrate solution (5 grams sodium thiosulfate pentahydrate per gallon) is required to neutralize 1 ppm of chlorine. If sodium sulfite is used, the neutralization ratio is about 6:1. However, sulfite solutions tend to be unstable and a treatment concentration of 2:1 should be used to provide for a safety margin.

Filtration through activated carbon is another commonly used dechlorination method. Activated carbon chemically reacts with both chlorine and chloramines, converting them into innocuous amounts of carbon dioxide and ammonium salts. Activated carbon filters will reliably reduce the free and bound chlorine concentration down to a few μg/L, but 100% dechlorination almost always requires supplemental thiosulfite additions. In the experience of the author, a flow-through filter containing 12 ft^3 of granular activated carbon will reliably dechlorinate about 25 gpm of water containing a total chlorine residual of 0.05 ppm for a period of about one year before needing replacement.

The fish toxicity of chlorine is high. At concentrations of 0.1–0.3 ppm, chlorine will kill most commercially important species fairly quickly at any pH. As mentioned, the OCl^- form is less toxic than HOCl but the practical importance of this is slight because little OCl^- is present over the pH range 6–8 typical of fish rearing conditions. Environmental conditions such as temperature and DO also influence chlorine toxicity but again, the practical importance of any protection gained is slight. To adequately protect warm- and coldwater fish from gill damage, chronic exposure to free chlorine should not exceed 3–5 μg/L. For short periods (up to 30 min), exposures as high as 0.05 mg/L can usually be tolerated.

Dissolved Oxygen

The need to provide an adequate amount of dissolved oxygen to fish in intensive culture is self evident. DO concentrations that are too low lead to serious adverse effects on health including anorexia, respiratory stress, tissue hypoxia, unconsciousness, and eventually death. However, high DO concentrations can be expensive to maintain. The challenge in aquaculture is to balance the economic benefits to health, feed conversion, and growth that an increased DO will provide against the cost of the aeration required to provide it.

Oxygen enters the water by passive diffusion from the atmosphere—a process driven by the difference between the partial pressures of oxygen in the air and in the water. At sea level, the oxygen in the atmosphere exerts a partial pressure (P_{O_2}) of about 157 mmHg (out of the total air pressure of 760 mmHg). The oxygen in water equilibrated with air exerts approximately the same partial pressure (157 mmHg). The P_{O_2} in the blood and tissues of fish is somewhat lower than this. However, a P_{O_2} of only 90 mmHg will saturate the Hb of coldwater fishes. Under normal conditions, there is always a substantial DO gradient between the water and the blood circulating through the gills.

The maximum amount of oxygen that will dissolve in water is a function of several variables including altitude, temperature, and salinity. For freshwater at sea level, and over the temperature range of interest in fish culture (0–30°C; 32–86°F), oxygen solubility can be calculated from the simplified equation developed by Soderberg (1995)

$$\text{DO (mg/L)} = 125.9/T^{0.625}$$

where T is water temperature (°F).

By convention, the DO is usually expressed in terms of its concentration by weight (mg/L) rather than by volume or partial pressure. As a first approximation, these terms can be interchanged over the range of conditions of interest in fish culture by the following conversion factors: 10 mg/L = 7.0 ml/L = 142 mmHg. Because fish culture operations may be conducted at higher elevations or in seawater in which oxygen is less soluble, DO levels are sometimes also expressed in terms of percent saturation rather than as mg/L (see Table 3.4).

Although a DO concentration equivalent to a minimum of 60% of saturation will provide sufficient O_2 tension to completely oxygenate the hemoglobin of salmonids (Klontz 1993), this provides no safety margin. Overall health and physio-

TABLE 3.4. Minimum dissolved oxygen (DO) concentrations recommended to protect the health and physiological condition of cold- and warmwater fishes during rearing.

Temperature		Oxygen Saturation	Minimum DO Level Required	
°C	°F	mg/L	mg/L	% saturation
5	41	12.8	9.1	71
10	50	11.3	8.8	78
15	59	10.2	8.3	81
20	68	9.2	7.8	85
25	77	8.2	7.4	90
30	86	7.5	6.9	92

Source: Data compiled from Wedemeyer et al. (1976)

logical condition are usually best if the DO is kept closer to saturation. If feeding is unrestricted, it is possible that almost any reduction in DO below air saturation[8] has at least minor effects on growth or other physiological function. Lower DO limits required to sustain life are well documented but of minor interest in well-run intensive culture systems. Examples are: tilapia, >0.6 mg/L; carp, channel catfish >0.8 mg/L; rainbow trout, coho salmon, >2 mg/L depending on the water temperature. Of greater interest are the limits of DO below which fish health problems due to hypoxia begin—6 mg/L at sea level for salmonids and 4 mg/L for warmwater species. As mentioned, a safety margin is needed to allow for temporary increases in O_2 consumption due to increased swimming activity, overfeeding, and CO_2 increases. Widespread agreement exists that maintaining the minimum DO level of 7 and 5 mg/L respectively is preferable and certainly more prudent.

Considerably less agreement exists on the upper DO limits that should be maintained during intensive rearing. Pure oxygen aeration to maintain supersaturated DO levels is increasingly being used to increase production by increasing the carrying capacity of the water available. Fish biomass can thus be increased without the need for facility expansion or increased water use. At present, DO concentrations as high as 15–20 mg/L are being used in O_2 supplemented intensive culture. Problems with gas bubble disease (GBD) can occur if the added O_2 increases the total gas pressure (TGP) to >100% of saturation. Edsal and Smith (1989) found no problem with GBD in salmonids reared at >180% O_2 saturation if N_2 saturation was reduced to <80%, thus maintaining the TGP no higher than 105–110%. In addition, problems with CO_2 can occur. As a first approximation, salmonids produce about 1.4 mg/L CO_2 per mg/L O_2 consumed. To keep the CO_2 <20 mg/L, the DO should be limited to about 15 mg/L (Watten et al. 1991). However, growth is not necessarily improved at this DO level and hematocrits and hemoglobin tend to decline (discussed further in Chapter 4). There is presently no consensus on the safe upper limit for DO.

Recommendations for minimum DO levels (at the raceway or pond outlet) that will promote fish health and physiological quality as well as assure Hb-O_2 saturation are tabulated in Table 3.4 over the temperature range of interest in freshwater intensive fish culture. Percent saturation limits given apply to fish grown intensively in saltwater as well.

Hardness

The total hardness (expressed as mg/L $CaCO_3$) of a hatchery water supply is primarily a measure of the amounts of calcium and magnesium salts present.

8. About 11 mg O_2/L in 10°C freshwater at sea level. Dissolved salts reduce the solubility of oxygen but this effect is negligible over the range of salt concentrations freshwater fish can tolerate. DO concentrations in sea water are significantly lower than in freshwater.

Other divalent dissolved metals such as iron, copper, zinc, and lead can also add to total hardness. However, they are naturally present only in trace amounts in water suitable for fish culture so their contribution is usually minimal.

Hardness and alkalinity are probably the most general measure of water type and potential for biological productivity in common use. Like alkalinity, hardness is also commonly taken as a measure of buffering capacity. Soft water is usually acidic whereas hard water tends to be alkaline. In many cases, the total hardness and alkalinity values will be similar. However, acidic ground can be quite hard but have little or no alkalinity. Natural waters can be classified in terms of total hardness as follows (EPA 1986):

Soft, 0–75 mg/L (as $CaCO_3$)
Moderate, 75–150 mg/L
Hard, 150–300 mg/L
Very hard, 300 mg/L and above

Soft water is low in calcium and other minerals needed for fish health but this can be tolerated if dietary intake is sufficient. Within limits, harder water is the more beneficial for fish health because it provided needed calcium and reduces the osmotic work required to replace blood electrolytes continually lost in the copious urine produced by fish in freshwater. Toxicity problems with trace heavy metals and copper containing disease therapeutants will also be minimized in harder water (>150 mg/L). The data developed by Alabaster and Lloyd (1980) on levels of copper safe for rainbow trout as a function of water hardness still serve as a useful guideline.

Hardness (mg/L as $CaCO_3$)	Safe Cu level (mg/L)
10	0.001
50	0.006
100	0.01
300	0.28

Fish in hard water may be slightly less susceptible to infectious pancreatic necrosis virus and bacterial kidney disease, although the incidence may actually be inversely proportional to the ionic composition rather than to hardness per se (Fryer and Lannan 1993). Epizootic ulcerative syndrome (EUS), a disease affecting milkfish and other warmwater fishes cultured in Southeast Asia appears to be more severe in waters of low total hardness (Das and Das 1993).

As a guideline, water in the 50–200 mg/L range with a pH of 6.5–9 and alkalinity of 100–200 mg/L (as $CaCO_3$) is considered desirable for the intensive culture of both cold- and warmwater fishes.

Heavy Metals

Except near natural mineral deposits, heavy metals such as zinc, copper, mercury, and lead are normally found in surface water supplies only in trace amounts. For example, copper concentrations in surface waters supplying anadromous fish hatcheries in the northwestern United States average about 10 μg/L. Heavy metals are, however, widely used in industry and are often discharged as effluents in the form of soluble chloride and sulfate salts. Another potential source in hatcheries is leaching from galvanized or copper plumbing fixtures and pipes—particularly in soft water areas. Heavy metal ions are highly soluble and toxic in soft water, but they usually precipitate as insoluble carbonates or hydroxides in hard alkaline waters (>150 mg/L $CaCO_3$, pH 8) which greatly reduces their toxicity. Suspended silt and sediments will also reduce heavy metal toxicity by adsorbing the ions, thus effectively removing them from solution. However, the total amount of a particular heavy metal that can be precipitated and therefore rendered nontoxic by a given volume of hard water is limited and should not be relied on to protect fish health. Such physical and chemical variables as high temperature and dissolved CO_2 may act to further decrease the safety margin.

Dissolved zinc and copper are probably the most commonly encountered heavy metals in hatchery water supplies. Although these ions in drinking water are essentially nontoxic to humans except in very high concentrations, they are very toxic to aquatic life, especially in soft water (<75 ppm as $CaCO_3$). Acutely lethal concentrations usually fall between 0.1 and 1 ppm depending on contact time, water chemistry, and biological factors such as species, age, and previous exposure history. The toxic effects of zinc or copper are typically delayed for a day or two after exposure, at which time sudden mortalities will occur for no apparent reason. Water chemistry analysis at this time may show only normal background concentrations. Synergistic effects are also common. Dissolved zinc is more toxic in the presence of copper. As an example, the LC_{50} for zinc at DO of 5 mg/L is about 80% of its LC_{50} concentration at a DO of 10 mg/L. Low DO and an increase in water temperature increases the toxicity of heavy metals in general.

Naturally occurring copper is rarely present in amounts over 0.05 mg/L in the surface waters used for fish rearing but can occur naturally at this or higher levels in well or spring water. This amount of naturally occurring copper in spring water of low (21 mg/L) alkalinity has been found to be the reason for inhibited smolt development in fall (ocean type) chinook salmon (*O. tshawytscha*) (Beckman and Zaugg 1988). In the absence of natural sources, dissolved copper in hatcheries usually comes from exposed copper and brass alloys in valves, screens, and pipes; industrial effluents; and the use of copper sulfate sprays ($CuSO_4$) for controlling algae and aquatic weeds such as eurasian water milfoil.

In soft water (<75 mg/L as $CaCO_3$), a maximum copper concentration of 0.006 mg/L is suggested as the chronic exposure limit to protect the smoltification of juvenile anadromous salmonids. Copper concentrations up to 0.01 mg/L can be tolerated by nonanadromous coldwater fishes if the hardness is >100 mg/L. Problems with copper toxicity are less common in warmwater fish culture because these species are usually more resistant. However, because copper toxicity is affected by so many variable factors, a maximum chronic exposure level of 0.1 of the 96-hr LC_{50} value determined using the particular species and water in question is a prudent guideline to use.

Because copper is also toxic to microorganisms, it finds some use as a therapeutant to treat external parasite infestations and is used extensively for aquatic weed control. However, the safety margin is small and it is prudent to only use copper (either as $CuSO_4$ or chelated with triethanolamine) for fish disease control in hatcheries with hard (>150 mg/L) water supplies.

In the case of anadromous salmonids, trace heavy metal exposure during freshwater rearing can have major deleterious effects on smoltification and early marine survival. The developing gill ATPase enzyme system of parr and smolts is sensitive to levels of dissolved heavy metals that are within the maximum safe exposure limits for nonanadromous fishes. Lorz and McPherson (1976) first showed that chronic copper exposure at only 20–30 μg/L, partially or completely inactivates the gill ATPase system required for normal smolt development. The biological damage is not apparent unless the fish are transferred into seawater, at which time mortalities begin. For conservation hatcheries, a more insidious but equally devastating consequence is that normal migratory behavior may be suppressed. The fact that smolts are unable to migrate directly into the ocean increases their residence time in the river and estuary—with consequent mortality from predation and diseases such as vibriosis and viral erythrocytic necrosis.

In contrast to copper, sublethal zinc or cadmium exposure during rearing apparently does not adversely affect migratory behavior. However, if even very low levels of copper (10 μg/L) are simultaneously present, downstream migration is affected, and normal gill ATPase activity is suppressed. Freshwater cadmium concentrations of more than 4 μg/L also result in a dose-dependent mortality if exposed coho smolts are transferred directly into 30 ppt seawater (Lorz et al. 1978). However, if a 5-day freshwater recovery period is allowed before saltwater challenge, survival returns to normal. Unfortunately, few alterations in growth or in feeding behavior occur to provide a warning and the smolts may appear to be normal when released.

Nickel or chromium exposure at up to 5 mg/L for 96 hours in freshwater apparently does not compromise migratory behavior or capability for ocean survival but sublethal mercury exposure results in a dose-dependent seawater mortality (Lorz et al. 1978).

Less is known about the effects of low-level heavy metal exposure on the smoltification and migration of pink, chum, or fall (ocean type) chinook salmon, which have very short stream residence periods, or sockeye salmon, which immediately migrate to lakes for extended rearing. Exposing sockeye smolts to copper at 30 μg/L impaired hypoosmoregulatory performance, as revealed by a seawater challenge test, and resulted in some mortality (Davis and Shand 1978). Mortality and hatching of sockeye salmon eggs were not affected by continuous exposure to 5.7 μg/L cadmium during incubation (Servizi and Martens 1978). The 168 hour LC_{50} was 8 μg/L at the fry stage. For copper, the incipient lethal level during the egg-to-fry stage was within the range of 37 to 78 μg/L for sockeye salmon and 25 to 55 μg/L for pink salmon. Copper inhibits egg capsule softening, but hatching mortality occurs only at concentrations also lethal to eggs and alevins. Dissolved copper is concentrated by eggs, alevins, and fry in proportion to exposure concentrations. For pink salmon, mortalities of eyed eggs and fry occur when tissue copper concentrations reach levels of 105 and 7 mg/kg, respectively.

Inorganic mercury at concentrations of only 2.5 μg/L during egg incubation causes embryo abnormalities in developing sockeye and pink salmon eggs. Hatching success, fry mortality, and growth are less sensitive to mercury exposure than embryo malformation.

Safe exposure guidelines for a number of heavy metals are summarized in Table 3.1.

Hydrogen Sulfide

Hydrogen sulfide (H_2S) is a highly toxic soluble gas that can occur in hatchery water supplies either naturally or as the result of pollution. H_2S rarely occurs in surface waters because they are usually aerobic and any sulfides present are fairly oxidized to sulfates. However, groundwaters can have natural sulfide concentrations of up to about 10 mg/L, and concentrations near coal deposits and volcanic areas may be higher. Groundwaters with high sulfide levels often contain high concentrations of iron and carbon dioxide as well.

Hydrogen and other sulfides occasionally occur in surface waters used for hatcheries as pollution from paper mills, chemical plants, and oil refinery effluent discharges. However, the principal source of hydrogen sulfide in either groundwater or surface waters is the anaerobic decomposition of organic matter by soil bacteria (*Desulfovibrio* spp.). These microorganisms respire anaerobically using sulfate (SO_4^{-2}) as an electron acceptor instead of oxygen. Bottom waters in stratified lakes and reservoirs are often anaerobic and high concentrations of hydrogen sulfide produced in the sediments can diffuse into the water near the bottom. If these bottom waters are discharged, sulfides may enter hatcheries some distance downstream.

In seawater or freshwater net pen culture systems, an accumulation of decaying organic matter on the bottom can lead to hydrogen sulfide in the water column. If problems occur, the net pens can be moved to deeper water or to a location with increased currents to reduce the accumulation of organic matter on the bottom. However, H_2S in marine finfish culture is not often a problem. It can be a problem in intensive brackishwater shrimp production. Seawater muds produce generally higher sulfide concentrations because sulfate concentrations in seawater are higher, and shrimp reside near the bottom. Sulfide accumulation is also a potential problem in freshwater crawfish culture.

Significant amounts of sulfides are not normally produced in flow-through raceway systems. However, groundwaters and bottom waters discharged from reservoirs used for flow-through culture systems may contain toxic amounts of hydrogen sulfide. These waters are often also low in dissolved oxygen and the aeration required will simultaneously remove the hydrogen sulfide by oxidation and volatilization. Increasing the raceway water velocity to >3 cm/s (discussed in Chapter 4) will help reduce the amount of organic matter that accumulates on the bottom. Hydrogen sulfide should also not accumulate in raceway culture systems using water recirculation unless decaying organic matter has been allowed to accumulate because of inadequate solids removal.

Sulfide concentrations in warmwater pond fish culture may be reduced by aerating and circulating the water to minimize anaerobic conditions near the pond bottom. Practices normally used to maintain adequate dissolved oxygen (limiting the amount of organic matter present, aeration, and water circulation or exchange) will also help reduce hydrogen sulfide concentrations in ponds. Periodic draining and drying will facilitate air oxidation of sulfides in the mud and will enhance decomposition of organic matter as well.

Hydrogen sulfide readily penetrates the gill epithelium and exerts its toxic effects by blocking the ability of cells to use oxygen. The net result is hypoxia similar to the effect of cyanide or to an oxygen depletion. Fish exposed to lethal levels of dissolved hydrogen sulfide first show increased ventilation rates, then cessation of ventilation. Death follows within a few minutes. Toxicity increases as dissolved oxygen concentrations decrease. However, even oxygen-saturated water will not prevent mortality and hydrogen sulfide concentrations over 0.5 mg/L are acutely lethal to most adult fish regardless of species.

Over the range of water temperatures and pH values of interest in fish culture, hydrogen sulfide dissociates into an equilibrium mixture of HS^- and H^+, the proportions of which are determined by the temperature, pH, (and salinity).

$$H_2S \rightleftarrows HS^- + H^+$$

As in the case of ammonia, the un-ionized form (hydrogen sulfide) is thought to be much more toxic than the HS^- ion. Similarly, chemical analyses measures total sulfide present ($H_2S + S^-$) and the proportion of un-ionized (toxic) H_2S present must be calculated. For convenience, the calculated percentage of total sulfide ($H_2S + HS^-$) existing in the toxic H_2S form is tabulated here for the range of pH and water temperatures of interest in freshwater fish culture (values in seawater will be about 15% lower). As seen, the fraction of toxic H_2S present increases in colder and more acidic water:

	Temperature (°C)				
pH	10	15	20	25	30
5.0	>99	99	99	99	99
6.0	97	94	92	91	90
7.0	65	59	55	51	47
8.0	14	12	11	9	8
9.0	<1	1	1	1	1

Although much of the total sulfide will be converted to the less toxic HS^- at pH 7 and above, the protective effect is slight unless the amount of H_2S present is very small to begin with.

Hydrogen sulfide can be removed from water by aeration which volatilizes it to the atmosphere, but the rate of stripping is relatively slow because H_2S is highly soluble. However, the HS^- ion is unstable in water of neutral to basic pH, and can be rapidly oxidized to sulfate by aeration. If iron or manganese is present, sulfides will be removed by reaction to form metal sulfides. These compounds give anaerobic muds their characteristic black appearance.

Hydrogen sulfide is determined in water supplies by measuring total sulfide concentration by chemical analysis and then calculating the H_2S concentration from the pH, temperature, and salinity. For concentrations down to about 0.1 mg H_2S/L, the methylene blue colorimetric procedure is the most commonly used. Concentrations below 0.1 mg/L require more involved procedures and specialized equipment (Tucker 1993).

At present, the recommended maximum safe exposure level is 0.002 mg/L for fish and other aquatic life in natural waters (EPA 1986). However, some variation in toxicity does exist because of species differences and the varying effects of environmental conditions. For example, coldwater fish with high oxygen consumption rates, and the eggs and fry of most species, are quite sensitive and can be affected by H_2S levels as low as 0.001 mg/L. For this reason, a lower exposure level of 0.001 mg H_2S–S/L is recommended as more prudent for both warm- and coldwater fish under intensive culture conditions (Tucker 1993). To prevent costly production losses, any measurable concentration of H_2S should be treated as a potential problem and steps taken to remove it.

Nitrate, Nitrite

Although nitrites (NO_2^-) and nitrates (NO_3^-) may contaminate hatchery water supplies directly as a result of agricultural drainage or sewage treatment plant effluent discharges, the normal source is indirect—the microbial oxidation of ammonia produced by fish metabolism and the decomposition of feces and uneaten feed by *Nitrobacter* and *Nitrosomonas* spp.

$$NH_3 + O_2 \rightarrow NO_2^-$$
$$NO_2^- + O_2 \rightarrow NO_3^-$$

Nitrites can accumulate to toxic levels (above about 13 mg/L for catfish, 0.3 mg/L for salmonids) if ammonia is converted to nitrite by nitrifying bacteria in excess of the rate of oxidation to nitrate. Nitrites may also be formed by the aerobic decomposition of excess organic material (e.g., plant matter, excess feed) in a pond, or by the anaerobic decomposition of organic matter in the pond bottom. The latter is especially likely in warmwater fish culture. In cold water hatcheries, nitrite accumulation occurs primarily in malfunctioning reuse systems with biofilters employing nitrifying bacteria for ammonia removal.

Nitrite exposure may cause fish health problems such as gill hypertrophy, hyperplasia, and lamellar separation together with hemorrhages and necrotic lesions in the thymus (Wedemeyer and Yasutake 1978). Chronic sublethal nitrite exposure is also suspected of increasing susceptibility to infectious diseases, but data to document this are lacking. The principal fish health problem caused by nitrite exposure is methemoglobinemia, commonly termed *brown blood disease* (discussed later in this chapter). This physiological condition is named after the brown-colored methemoglobin formed when nitrite oxidizes the Fe^{+2} iron in the Hb molecule to the Fe^{+3} state. The oxidized form of Hb (methemoglobin) is unable to bind oxygen and carry it to the tissues.

The 96-h LC_{50} of nitrite is about 13 mg/L (NO_2-N) for channel catfish and 0.3 mg/L for rainbow trout when measured under controlled conditions (Russo and Thurston 1977). In practice however, the toxicity is strongly influenced by the other anions present (particularly chloride), and by the pH. Increasing the dissolved calcium or chloride ion concentration to 50 mg/L can increase the tolerance of salmonid fishes to nitrite by a factor of 50 (Wedemeyer and Yasutake 1978; Tomasso et al. 1980; Russo et al. 1981). These ions presumably compete with nitrite for transport across the gill epithelium and therefore decrease the amount of nitrite uptake. Decreasing the pH increases nitrite toxicity by converting some of the NO_2^- present to HNO_2. However, the pK_a for this reaction is low (3.14 at 25°C) and the practical effect is slight over the pH range of 6–8 typical of fish cultural conditions. A nitrite level of <0.1 mg/L (as NO_2^-) should be adequate to protect fish health under most water quality conditions.

Nitrate (NO_3^-) is commonly considered to be essentially nontoxic to fish. For example, chronic exposure to sodium nitrate at up to 200 mg/L has no effect on the growth of channel catfish (Colt and Armstrong 1981). The 96-h LC_{50} of sodium nitrate to salmonids and most other species is 1,000–3,000 mg/L NO_3^--N, comparable to the lethal concentration of NaCl (Colt and Armstrong 1981). Thus, high nitrate concentrations may simply cause nonspecific osmoregulatory failure similar to that caused by high salt concentrations in general. However, high nitrate exposure during egg incubation has been shown to have adverse effects on embryo development in salmonids (Kincheloe et al. 1979). In hatcheries, this may potentially occur when recycle systems with biofilters for ammonia oxidation are used or when water supplies have been contaminated by irrigation return flows from intensive agriculture operations. A maximum chronic exposure level of <1.0 mg/L is a suggested guideline, but under most circumstances the toxicity of nitrate to fish in intensive or extensive culture should not be a serious problem.

Supersaturation

Water supersaturated with air or other dissolved gas has long been a significant source of fish health problems, both in natural waters and in aquaculture facilities worldwide. Under normal conditions, the partial pressures (mm Hg) of the oxygen, nitrogen, argon and carbon dioxide dissolved in water from the air are in balance with the pressure exerted by these gases in the atmosphere—usually about 760 mm Hg at sea level. However, this balance can be easily altered by a variety of natural or artificial conditions, such as temperature changes or air leaks on the suction side of centrifugal pumps that cause the total pressure of one or more of the dissolved gases in the water to become greater than the atmospheric pressure. The blood and tissues of fish quickly reach equilibrium with the partial pressures of the dissolved gases present in the water and if the water becomes supersaturated, the blood and tissues will also become supersaturated. Unfortunately, the gases dissolved in the blood and tissues may also come out of solution, forming bubbles in the vascular system (embolisms) that can physically block circulation. Bubbles form in the vascular system by several mechanisms including turbulence as blood is pumped through the heart, reduction in blood pressure in the venous side of the circulation, and the physical phenomenon of bubble formation on surfaces (McDonough and Hemmingsen 1985). Depending on the extent, hemostasis, tissue necrosis, and death can quickly follow.

Normally, a membrane diffusion instrument such as one of the commercially available saturometers is used to measure supersaturation. A gas permeable membrane probe is placed in the water in question and degree of saturation (as a percentage of the total gas pressure) is calculated using the following basic formula:

$$TGP(\%) = (BP + \Delta P)/BP \times 100$$

where

TGP(%) = percent of total gas pressure
BP = barometric pressure
ΔP = differential gas pressure measured by the instrument.

A standard method for supersaturation analysis using membrane diffusion instruments has been developed by Colt (1989). Supplemental computer programs that calculate additional parameters such as the percent saturation of oxygen or nitrogen from the TGP(%) and DO concentration are also available (Dawson 1986). As a general rule, anything more than about 110% of air or nitrogen supersaturation (assuming $\Delta P > 0$) is likely to eventually cause some degree of GBD. If ΔP is zero or less, bubbles cannot form, even if one of the individual gases dissolved from the atmosphere (such as nitrogen) is supersaturated as commonly happens.

As mentioned, GBD usually occurs because the hatchery water supply has become supersaturated with air. However, GBD can also occur if one or more of the individual gases (normally either nitrogen or oxygen) become supersaturated. In pond culture systems, oxygen supersaturation can occur during algae blooms or because of the photosynthesis of aquatic plants. In hatcheries using well or spring water, nitrogen supersaturation is fairly common. Oxygen supersaturation is usually much less of a fish health problem because tissue metabolism tends to remove any oxygen bubbles that form.

Supersaturation in hatchery water supplies has several causes. Groundwater from springs and deep wells is a particularly common cause. Groundwater originates from rainfall that has percolated underground to considerable depths, sometimes forming subterranean rivers and lakes. This water can contain large amounts of dissolved air and is usually under great pressure. Soil microorganisms frequently utilize most of the dissolved oxygen, leaving mainly nitrogen and the carbon dioxide they have produced. When this water is brought to the surface by pumping (or naturally, as a spring), the pressure is released and the water becomes supersaturated with nitrogen, and to a lesser extent carbon dioxide because of its higher solubility.

Another common, but easily overlooked, cause of supersaturation is air drawn through small leaks into the suction side of centrifugal water pumps and then forced into solution under pressure. Gas supersaturation occurs as the pumped water is discharged into fish rearing units at atmospheric pressure. Such air leaks can be difficult to detect because visible bubbles may not be present in the discharged water, especially in seawater hatchery systems (Kils 1977). A similar situation can arise in a gravity flow water supply if air bubbles are entrained into water flowing down a standpipe drain. This can easily happen if the standpipe is not completely submerged. The problem can be corrected by sizing the diameter of the standpipe to maintain a head of water above the intake, thus preventing air entrainment. For experimental purposes, it is possible to take advantage of air entrainment to achieve a controlled amount of supersaturation. For example, air drawn in through a needle valve installed on the suction side of a centrifugal pump can be regulated to attain supersaturation levels as high as 140%.

Another cause of supersaturation problems in hatcheries is heated water. Because gas solubility decreases as the temperature rises, cold water at air equilibrium will become supersaturated if it is warmed even a few degrees. As a rule of thumb, each increase of 1°C increases supersaturation by about 2% (Marking

1987). Thus, warming air-saturated water by only 5°C will result in 110% supersaturation. For example, cold snowmelt, which is normally saturated with air, can became supersaturated after flowing into warmer stream water and cause GBD in sensitive salmonid fry. In a related situation, dissolved gases released when pond water begins to freeze can be forced into the remaining unfrozen water under pressure causing supersaturation and GBD (Mathias and Barcia 1985).

Finally, supersaturation in hatchery water supplies can be caused by air entrained into water flowing over the spillways of dams or over waterfalls (Colt 1984). In either case, entrained air is dissolved under pressure as the falling water plunges below the surface. The water can become supersaturated to the extent that resident fish are killed by GBD. If the water falls onto rocks instead of into a plunge basin, it will tend to become air equilibrated and be less lethal to fish.

Fish may or may not detect and avoid supersaturated water but this ability varies with species, degree of supersaturation, and temperature. Rainbow trout and most of the Pacific salmon can generally detect and avoid 125% and 145% saturation respectively, but neither will avoid 115% (Stevens et al. 1980). Anadromous steelhead trout do not avoid even highly supersaturated water. In hatchery situations the question of avoidance is usually irrelevant. However, deeper rearing units may offer some protection because the greater average depth of the fish in the water column will provide a degree of hydrostatic compensation (Colt et al. 1991).

A wide variety of degassing devices that can reduce supersaturation to acceptable levels have been developed. Most simply provide increased turbulence and increase the area of the air-water interface to speed up the rate at which supersaturated gases in the water can reach equilibration with atmospheric pressure. Early methods for degassing supersaturated water employed mechanical agitation to increase turbulence and surface area and to achieve aeration. Many of these methods, spraying or cascading, formation of thin layers to increase the surface area, or air stripping the excess gas by aeration are still used. Although these simpler approaches are effective in reducing supersaturation to 104–105% or less, none will eliminate it entirely. To achieve levels of 100% or less, more technically advanced equipment such as packed columns and vacuum degassing units must be used. Packed column degassing units typically consist of a length of 4–6 inch plastic pipe filled with about 4 ft^3 of commercially available 2 inch diameter plastic rings (e.g., Norton rings) to increase the surface area and turbulence of water flowing through the column (Owsley 1981). The columns are used vertically with gravity flow. Packed-column degassing is one of the most practical methods yet developed for air equilibrating supersaturated water. Unlike past techniques, correctly designed packed columns can decrease supersaturation to nearly 100% (Owsley 1981).

If gas saturation must be reduced to $<100\%$, some degree of vacuum has to be applied to the supersaturated water. In most cases, reducing the atmospheric pressure by only 10–20 mm of Hg is all that is required to release the residual supersaturated gases (Fuss 1986). In a comparison of packed columns and packed columns with vacuum degassing, packed columns alone decreased supersaturation from 133% to about 104%. Packed columns with an applied vacuum easily decreased supersaturation to less than 100% (Marking et al. 1983). Unfortunately, vacuum degassing will also reduce the DO concentration, whereas packed columns used alone will normally increase it. Oxygen injection into packed column vacuum degassing units is an inexpensive technique that can be used to correct both supersaturation and DO deficiency problems (Marking 1987).

The currently recommended maximum safe exposure level to prevent GBD is 110% supersaturation (EPA 1986). However, interpreting this figure requires the user to take into account the fact that sensitivity to GBD varies with water depth and between fish species. Thus, the 110% standard is too high to adequately protect fish chronically exposed to supersaturation in shallow water raceways and too low for fish in ponds or natural waters deeper than 1 m (Alderdice and Jensen 1985). A more appropriate safe chronic exposure level for salmonid fish in deep rivers and lakes is 115%. In hatchery ponds and raceways, GBD can occur in juvenile Pacific salmon chronically exposed to supersaturation levels above 104–105%. In more sensitive species such as Atlantic salmon and lake trout, GBD has occurred in juveniles exposured to 102% (Marking 1987; Krise and Meade 1988). For most salmonids in intensive culture, supersaturation should not be allowed to exceed 103% to protect eggs, fry, and fingerlings, and 105% to protect fish up to smolt size.

Temperature Limits

The health of fish in intensive culture is affected by both variations and extremes in the temperature of the water in which they live. Each species has a water temperature range within which it thrives. Sustained temperatures outside this range, or rapid temperature changes within this range, represent stressful or lethal environmental conditions.

Warming the water increases the toxicity of any dissolved contaminants, generally promotes the growth and invasiveness of fish pathogens, decreases the DO concentration, and increases oxygen consumption by increasing body temperature and thus the metabolic rate. As calculated by Klontz (1993), warming a unit volume of water at 9°C to a temperature of 15°C decreases the available DO by about 13% while causing the metabolic rate of a 100 g rainbow trout living in this water to increase by 68% and its ammonia excretion to increase by 99%. Cooling the water lowers body temperature, slows the immune response, and reduces feeding, activity, and growth. The end result is that water temperature probably has a greater effect on fish health and physiological condition than any other environmental variable with the possible exception of dissolved oxygen.

Salmonids and other coldwater species economically important to intensive fish culture have a somewhat narrow thermal tolerance range (termed *stenother-*

mal) and are rather sensitive to (rapid) water temperature changes even within this range. Optimum water temperatures for spawning behavior, egg incubation, and smolt development are generally lower than the optimum temperatures for growth. Tropical fish such as tilapia are also stenothermal and usually quite cold intolerant. Warmwater species such as channel catfish are generally *eurythermal*; that is, they can tolerate a broader range of water temperature. When either stenothermal or eurythermal species are reared extensively in ponds however, it becomes more difficult to evaluate temperature effects because changes in water temperature affect not only the fish, but the ecology of the pond as well. Every plant and animal inhabiting the pond grows and reproduces within a definite temperature range. A change of a few degrees can cause populations of some aquatic organisms to decline while others greatly increase. The resulting differences in pond ecology can make it difficult to interpret the effects of water temperature on the fish themselves.

The maximum and minimum temperatures that fish can tolerate is genetically determined but is also influenced to some extent by variables such as length of time allowed for acclimation, DO concentration, and on the amount and kind of dissolved ions that may be present. For this reason, thermal death point temperatures are not fixed values but can vary by several degrees. Thermal death points are usually of limited interest in hatchery work. Approximate values for a few economically important species: rainbow trout, 0 and 26°C; carp, 0 and 30°C; channel catfish, 4 and 35°C; tilapia spp., 10 and 38°C.

The physiological mechanisms responsible for death at high water temperatures are not completely understood but several factors are probably contributory. The oxygen consumption of rainbow trout for example, rises by an order of magnitude as the water is warmed from 0°C to 26°C, the approximate lower and upper lethal temperature limits for this fish. At the same time, the DO declines by at least half because of reduced solubility. Blood O_2 levels steadily decline and oxygen transport to the tissues is reduced (Houston and Koss 1984). Ability to maintain energy reserves (whole body lipid content) and serum electrolyte concentrations also declines as the water is heated. Osmoregulatory failure may be the ultimate cause of death. Sodium, magnesium, and calcium ions added to the water provide a degree of protection to most freshwater fish against the effects of high temperature.

Death due to low water temperatures likely involves physiological mechanisms similar to those responsible for heat death. For example, tilapia, a widely cultured tropical fish, show several clinical signs of distress as the water temperature is slowly lowered below their optimum temperature range of 20–30°C. At about 18°C, reproductive behavior begins to be affected. Feeding and growth cease at about 15°C and the fish become inactive and disoriented. Below about 10°C, most tilapia species become comatose suffering what is commonly termed *chill coma*. During chill coma, serum total protein, sodium and chloride ion concentrations, and plasma osmotic pressure steadily decline. Death occurs due to osmoregulatory collapse and renal failure. Cold water survival can be improved somewhat by adding electrolytes to the water or by holding the tilapia in seawater diluted to a salinity of 5–10 ppt (Sun et al. 1992).

Tolerance to Temperature Changes. In most fish culture situations, the physiological effects of temperature changes are of more practical importance than lethal temperature extremes. Temperature shock syndrome caused by changes in water temperature that are too rapid is a well-recognized problem that experienced fish culturists take precautions to avoid.

Although gradual water temperature changes of 10°C or less over a period of hours seldom cause fish health problems, 10°C changes over a period of minutes cause a stress response in salmonids that requires about 24 hours for recovery (Wedemeyer 1973). Allanson et al. (1971) were among the first to show that tilapia could be transferred directly from 25°C water to 30°C with no problems but if moved directly into 15°C water, they did not recover fully even after 20 days. Rapid temperature changes will cause lethal temperature shock in other species as well. For example, threadfin shad acclimated to 15°C and then chilled suddenly to 5°C are well-known to survive less than 24 hours. Rapid temperature changes of 10°C or more can also activate latent infections. Fish sub clinically infected with *Aeromonas hydrophila* are much more sensitive to temperature changes than are noninfected fish and activated infections are often the cause of the temperature shock mortalities. In the case of incubating salmonid eggs and yolk sac fry, rapid temperature changes of only a few degrees may cause coagulated yolk or white-spot disease (Wedemeyer and Goodyear 1984).

The question of how quickly water temperatures (and thus body temperature) can be changed without exceeding physiological tolerance limits has yet to be resolved, partly because it varies with the previous temperature history of the fish in question. Nickum (1966) originally showed that a variety of economically important cold- and warmwater species could survive sudden changes of less than 11°C. Today, however, most culturists recommend that fish should be allowed to gradually acclimate to warmer or colder water over a period of two or more hours if the temperature difference is >10°C.

Effects on Growth. In most fish culture situations, optimum water temperatures for health and growth are of more interest than upper or lower lethal limits. Temperature affects the rate of growth and development by affecting a variety of metabolic processes including respiration, feeding, and digestion. Any divergence in the normal range of these processes may alter the optimal range for health and growth. In addition, the temperatures required for optimum growth may or may not coincide with those required for optimum fish health. This must usually be worked out empirically for each hatchery situation.

Temperature guidelines for the growth of channel catfish are: egg incubation and fry feeding, 26–28°C; digestion, 27–29°C; growth, 28–30°C; and overall best survival, 28–30°C. Problems with bacterial diseases and channel catfish virus disease (CCVD) are generally worse at water temperatures >28°C. Channel catfish avoid >35°C or <5°C water (Tucker and Robinson 1990).

The growth of rainbow trout is optimized at 16–17°C although the efficiency with which they assimilate food increases up to a maximum of about 85% at 20°C. The parr–smolt transformation of the anadromous form of rainbow trout (steelhead) is inhibited at >13°C (discussed in Chapter 4).

Effects on Smolt Development. Elevated water temperatures are sometimes used to accelerate growth and shorten the normal time needed to produce smolts. However, such artificial temperature regimes must be used with care because they can influence the smolting process itself as well as growth. Rearing temperature has a strong influence on the pattern of hypoosmoregulatory and gill ATPase activity development during the parr–smolt transformation. In certain species, elevated water temperatures not only accelerate the onset of smolting but also hasten the process of parr-reversion (desmoltification) so that the duration of the smolting period is shortened. For example, Zaugg and McLain (1976) were the first to show that the rise in gill ATPase activity of juvenile coho salmon is retarded at 6°C, normal at 10°C, and precocious at 20°C. Temperatures up to 15°C can be safely used to accelerate physiological development so that juvenile coho salmon can be introduced to seawater in their first year.

Not all anadromous salmonids are amendable to temperature acceleration. Zaugg and Wagner (1973) first showed that steelhead trout are particularly sensitive to the elevated rearing temperature required to accelerate growth. Smolting can be almost completely inhibited at 13°C and above. In contrast, Atlantic salmon apparently smolt at temperatures as high as 15°C. However, wild runs of Atlantic salmon show their greatest downstream migratory behavior as the temperature rises to 10°C. Outmigration has usually been completed before the water warms to 15°C or more.

Effects on Disease Resistance. Within genetically defined limits, water temperature changes alter the metabolic rates of both fish and their microbial pathogens. The immune response of both cold- and warmwater fish are enhanced by temperature increases. In contrast, low temperatures suppress the immune response of warmwater fish while only delaying and decreasing its intensity in coldwater fish. Thus it is not surprising that temperature plays such a major role in infectious disease processes. The severity of infectious diseases caused by most gram-negative bacteria is greater at elevated water temperatures. The mean time from infection to death is shortened and a greater total mortality also occurs. However, warmer water temperatures do not solely benefit pathogens. Their metabolism and rate of cell division may be increased at higher temperatures but the kinetics of host defense processes such as tissue repair are also accelerated. For example, epithelial cell renewal time in the small intestine of coho salmon is thought to play a role in resistance to vibriosis (Groberg et al. 1983). Cell renewal time is >35 days at 5°C but only 14 days at 18°C.

Some pathogens, such as *Flexibacter psychrophilus* (bacterial coldwater disease of slamonids) and *Edwardsiella ictaluri* (enteric septicemia of catfish), have cool optimum temperatures relative to their hosts. The mortality rate may decrease as the water is cooled but death will still occur after a more prolonged incubation period. Many fish pathogens are also psychrophilic (e.g., *Vibrio anguillarum*) and thus remain capable of relatively rapid cell division at cold water temperatures that depress the host immune protection system. For example, metabolism and cell division of the vibrio bacterium above 15°C are so rapid that host defenses are easily overcome. At water temperatures <12°C, the growth rate of the microbe is suppressed such that immune protection system of the fish may more often prevail. At the intermediate temperatures of 12–15°C, the outcome may be determined by more ancillary factors such as the fishes' nutritional condition, degree of physiological stress, and integrity of the mucous barrier. Numbers of invading microbial cells are probably also more important in the intermediate temperature range (Groberg et al. 1983). To provide fish culturists with predictable results on the control of infectious diseases under varying environmental temperatures, knowledge of the potential for repression of immune protection at reduced temperatures is required together with knowledge of the ability of each pathogen to grow and induce mortality at these temperature.

While the optimal temperature ranges for most fishes of economic importance have been well documented, the ranges that potentiate fish diseases have received less attention. For example, handling channel catfish fingerlings during hot weather (32–38°C air temperature) often predisposes to infections by myxobacteria and fungi. Under such conditions, water temperatures will temporarily far exceed optimal limits and the fish will prove extremely susceptible to disease. If handling is necessary, adding salt (NaCl) to the water at 0.5–1.0% will help increase survival and reduce rates of infection. Outbreaks of bacterial diseases in channel catfish in pond culture occur most frequently during July and August, the two hottest months of the year. Even though low oxygen levels and metabolic byproducts from the decomposition of organic matter are probably the most important factors, the added stress of higher than optimal water temperatures and the limiting effect of such high temperatures on the oxygen saturation levels of the water cannot be ignored (Wedemeyer et al. 1976). Very obvious effects on health and disease resistance can also be noted when fish are held for extended periods under temperature regimes below their optimum. Tilapia and channel catfish held at 17°C during winter months are more susceptible to infection by *Saprolegnia* and *Chilodenella*. Routine chemical treatments readily control these problems on channel catfish but not on tilapia. Both tilapia diseases are amenable to treatment at 25°C. A summary of guidelines on water temperatures and fish diseases is given in Table 3.5.

TABLE 3.5. Influence of water temperature on bacterial, parasite, and viral diseases of fish in intensive culture (data compiled from Ahne and Wolf 1980; Reichenbach-Klinke 1981).

Pathogen Type	Disease Problems Less Severe: Under (°C)	Over (°C)
Bacteria		
Aeromonas salmonicida	3.2	25
Aeromonas hydrophila	10	—
Flexibacteria sp.	3.5	20
Renibacterium salmoninarum	5	—
Carp erythrodermatits	10	30
Parasite		
Icthyobodo necatrix	2	30
Ceratomyxa shasta	3.9	24
Myxobolus cerabralis	3	25
Ichthyophthirius sp.	3	27
Dactylogyrus sp.	5	28
Lernaea sp.	10	30
Virus		
Viral hemorrhagic septicemia (VHS)	2	20
Infectious pancreatic necrosis (IPN)	4	26
Infectious hematopoietic necrosis (IHN)	4	20
Channel catfish virus (CCV)	20	30

In most fish culture situations, optimum water temperatures for growth (sometimes termed the *standard environmental temperature*) rather than lethal maximums or minimums are of the most interest. Unfortunately, the water temperatures required for optimum growth may or may not coincide with those required for optimum disease prevention. So many factors influence the effects of temperature on fish pathogens that only broad recommendations can be given. However, the general guidelines for bacterial, fungal, and viral diseases given in Table 3.5 are useful as a first approximation. Each hatchery situation is different and specifics must usually be worked out empirically for each disease.

Upper water temperature limits recommended for the culture of selected economically important fishes are summarized in Table 3.6.

Total Dissolved Solids, Salinity

Total dissolved solids (TDS) is the collective term for the total concentration (in mg/L) of all the inorganic mineral salts dissolved in water. Salinity, an oceanographic term, also describes the total dissolved salt concentration of the water but is usually expressed as grams/kg or parts per thousand (ppt or ‰). Salinity and

TABLE 3.6. Suggested upper water temperature limits to protect the health and physiological condition of selected cold- and warmwater fishes during rearing (data compiled from Huner and Dupree 1984; Klontz 1993; and the experience of the author).

Fish Species	Egg Incubation		Rearing	
	°C	°F	°C	°F
Atlantic salmon	5	41	17	63
Rainbow trout	9	48	17	62
Pacific salmon	10	50	12	54
Steelhead trout	10	50	12	54
Striped bass	20	68	32	90
Channel catfish	27	81	30	89

TDS are closely related terms and often used interchangeably. For many purposes the two concepts are equivalent (EPA 1986). The principal dissolved minerals making up the TDS or salinity are carbonates, chlorides, sulfates, nitrates, and salts of sodium, potassium, calcium, and magnesium. Soluble organic materials may also be present in small amounts. Together with natural sources, dissolved minerals in industrial effluents and agricultural drainwater can also add to the salinity or TDS load of hatchery water supplies.

Strict guidelines for salinity/TDS have not been developed, but most natural waters that support fish populations have a salinity of 0.1–1 ppt (10–1,000 ppm TDS). The toxicity of the individual mineral salts making up the salinity or TDS is usually not a major problem in fish culture—partly because many of the dissolved ions exert antagonistic effects upon one another. However, salinity does increase the osmotic pressure of the water and mortality from osmoregulatory failure will result if the salinity becomes too high. Adult rainbow trout can be acclimated to a salinity of 30 ppt but the maximum salinity at which most noneuryhaline freshwater fish can osmoregulate successfully is about 10 ppt. Channel catfish can tolerate salinities as high as 8 ppt but this figure is the upper limit for survival, not a guideline for good fish health and physiological condition. Channel catfish do best at salinities of <3 ppt.

Total Suspended Solids, Turbidity

A certain amount of turbidity in hatchery water supplies is quite common due to suspended clay and soil sediments from natural siltation, or from construction, mining, and logging. In ponds and self-cleaning raceways[9], fecal solids and uneaten feed also contribute. Suspended particulate matter can suffocate developing

9. In self-cleaning raceways, water velocities >3 cm/sec are used to prevent particles of uneaten food and such from settling out (discussed in Chapter 4).

eggs during incubation, interfere with sight feeding, and physically abrade or coat the gills. Physiological consequences include stress and reduced disease resistance. For example, juvenile salmonids chronically exposed to high concentrations of suspended soil particles, clay, and volcanic ash suffer a physiological stress response and reduced resistance to subsequent infections (Redding and Schreck 1987).

Total suspended solids (TSS) and turbidity concentrations favoring optimum fish health are not known, but 80–100 ppm TSS is considered a reasonable chronic exposure level that will protect salmonids and other sensitive species against gill damage. Unfortunately, safe exposure guidelines have not been developed for suspended solids with irregular sharp edges that can cause physical gill damage such as fresh volcanic ash and certain pollens (Klontz 1993).

Guidelines are also not well defined for turbidity. For hatcheries using ultraviolet (UV) water treatment systems, sand filtration to reduce turbidity to <20 NTU (Natelson turbidity units) is recommended to assure adequate UV penetration (discussed in Chapter 6). For salmonid fishes, limiting turbidity increases to <20 NTU over ambient levels will minimize interference with sight feeding.

Summary

Guidelines for the overall water quality conditions that should be maintained to promote the health and physiological condition of fish in intensive culture are summarized in Tables 3.1–3.6. Fish can often physiologically compensate for the alterations in water quality that occur during rearing (e.g., oxygen depletions or temperature increases) when they occur individually. When several otherwise sublethal alterations occur at the same time (e.g., oxygen depletions and temperature increases), they may become debilitating and life threatening. Thus, the recommendations in Tables 3.1–3.6 are not meant to be taken as individual limits of physiological tolerance but rather as the overall water quality environment that will promote the general health and disease resistance of fish in intensive culture.

ASSOCIATED FISH HEALTH PROBLEMS

Gas Bubble Disease (Gas Bubble Trauma)

Under normal conditions, the partial pressures (mm Hg) of the gases (oxygen, nitrogen, argon, carbon dioxide) dissolved in water from the air are in balance with the pressure exerted by these gases in the atmosphere—usually about 760 mm Hg at sea level. However, if the water becomes supersaturated the total pressure of one or more of the dissolved gases in the water (commonly indicated by the symbol ΔP) will become greater than the atmospheric pressure. The blood and tissues of fish living in supersaturated water quickly reach equilibrium with the partial pressures of the dissolved gases present and if their total pressure (ΔP) is too much greater than atmospheric, vascular embolisms will form. Gas bubbles characteristically develop first in the circulatory system and are rapidly carried to the skin, the fins, and the mouth where they are easily seen. Bubbles that form in the heart or other vital organs normally cause death. Visible gas bubbles will also form in the yolk sac of fry, causing them to float upside down at the surface. Most of the fry that survive will eventually develop white spots of denatured protein in the yolk material and die of coagulated yolk disease at the swim-up stage.

The fish health problem resulting from supersaturation is commonly referred to as gas bubble disease (GBD) or gas bubble trauma. Gas bubble disease in fish differs from the related bends or decompression sickness in human divers because there does not have to be any sudden reduction in external pressure for bubbles to form. Bubbles probably form in the vascular systems of fish because of turbulence in the heart, reduction in blood pressure on the venous side of the circulation, and physical bubble formation on surfaces (McDonough and Hemmingsen 1985). If ΔP is zero or less, bubbles cannot form, even if one of the individual gases dissolved from the atmosphere (such as nitrogen) is supersaturated as commonly happens. Supersaturation causes mortality from GBD under a wide variety of rearing conditions. As mentioned, GBD usually occurs because the hatchery water supply has become supersaturated with air. However, GBD can also occur if one or more of the individual gases (normally either nitrogen or oxygen) have become supersaturated. Oxygen supersaturation is usually much less of a problem than nitrogen because tissue metabolism tends to remove any oxygen bubbles that form. As a general rule, anything more than about 110% of air or nitrogen supersaturation (assuming $\Delta P > 0$) will eventually cause GBD.

The currently recommended maximum chronic safe exposure level to prevent GBD is 110% supersaturation. This figure is an average and is based on the fact that sensitivity to GBD varies with water depth and between fish species. Thus, the 110% standard is too high to adequately protect fish chronically exposed to supersaturation in shallow water in hatcheries and too low for fish in natural waters deeper than 1 m (Alderdice and Jensen 1985). A more appropriate safe chronic exposure level for deep rivers and lakes is 115%, whereas in hatcheries, chronic exposure to supersaturation levels above 104–105% in Pacific salmon and 102% in more sensitive species such as Atlantic salmon and lake trout (*Salvelinus namaycush*) can result in GBD (Marking 1987; Krise and Meade 1988). For most salmonids in intensive culture, supersaturation should not be allowed to exceed 103% to protect eggs, fry, and fingerlings, and 105% to protect fish up to smolt size (Owsley 1981; Alderdice and Jensen 1985).

As mentioned earlier, some fish used in aquaculture can detect and avoid supersaturated water but this ability varies with species, degree of supersaturation, and temperature. Rainbow trout and most of the Pacific salmon can generally detect and avoid 125% and 145% saturation respectively, but neither will avoid

115% (Stevens et al. 1980). Anadromous steelhead trout do not avoid even highly supersaturated water.

Fish in distress from GBD can recover if they are held under increased hydrostatic pressure such as in water deeper than 1 m, or if the water temperature is lowered (Knittel et al. 1980). Therefore, it is of interest to consider the effects of pressure on GBD in more detail. As a first approximation, gas supersaturation decreases about 10% per meter of water depth because of the increased hydrostatic pressure (EPA 1986). Thus, if a unit volume of water that is 130% saturated at the surface is submerged to a depth of 3 m, the saturation would decrease to 100% because of the increased hydrostatic pressure. Similarly, fish at a depth of 3 m in water 130% saturated with air at the surface would experience 100% saturation if they remained at that depth. Fish equilibrated to 100% saturated water at a depth of 1 m would be subjected to 110% supersaturation if they were rapidly brought to the surface. In contrast to the GBD problems caused by rapid supersaturation increases, fish seem to adjust quite easily to rapid decreases in gas supersaturation such as might occur during ponding or stocking operations.

Methemoglobinemia (Brown Blood Disease)

Chronic nitrite exposure may cause fish health problems such as gill hypertrophy, hyperplasia, and lamellar separation together with hemorrhages and necrotic lesions in the thymus. However, the principal fish health problem caused by nitrite exposure is brown blood disease (methemoglobinemia) named after the brown-colored methemoglobin formed when nitrite oxidizes the Fe^{+2} iron in the Hb molecule to the Fe^{+3} state. The oxidized form of Hb (methemoglobin, M-Hb) is unable to bind oxygen and transport it to the tissues. Normal blood M-Hb concentrations average 1–3% and blood concentrations >10% are considered undesirable. After more than about 50% of the Hb is oxidized to M-Hb, death from hypoxia becomes a strong possibility, especially if the DO is also allowed to decline. Clinical signs of methemoglobinemia or brown blood disease begin at M-Hb concentrations of 25–50%. Typical signs include lethargy, crowding near water inlets or aeration equipment, a characteristic chocolate-brown color of the blood, and a sharp rise in the daily mortality rate (Bowser 1984). Because respiratory efficiency is reduced by methemoglobinemia, tolerance to even minor oxygen depletions will be correspondingly reduced (Bowser et al. 1983). Supplemental aeration may be temporarily required to support affected fish. Brown blood disease has been reported in both salmonid and channel catfish culture. Susceptibility to the disease varies somewhat with genetic strain, presumably due to genetic differences in the erythrocyte M-Hb reductase enzyme system, which reduces M-Hb back to Hb (Tomasso et al. 1980).

The severity of brown blood disease is closely tied to water quality. Problems typically occur under rearing conditions in which high nitrite concentrations develop in water with low concentrations of dissolved chloride. Mortality increases with decreasing chloride pH, alkalinity, and DO, and increasing nitrite concentrations. However, the disease has been reported in channel catfish in ponds with nitrite concentrations as low as 0.5 ppm. For this reason, measuring the dissolved nitrite concentration alone is not adequate to predict outbreaks of brown blood disease (Schwedler and Tucker 1983).

Brown blood disease is best prevented through good fish cultural procedures rather than treated after it occurs. However, if the disease does occur, the standard treatment method is to apply sodium chloride or calcium chloride to the ponds involved (Bowser 1984). Calcium chloride is more effective than sodium chloride in treating methemoglobinemia in hatchery salmonids (Wedemeyer and Yasutake 1978). However this difference has not been demonstrated in channel catfish pond culture (Tomasso et al. 1980). The physiological basis for either salt as a therapeutant is that Cl^- competes with nitrite for uptake sites on the gill epithelial membrane. Thus, salt treatment minimizes or prevents further nitrite uptake and M-Hb formation. The normal erythrocytic M-Hb reductase system then slowly clears the M-Hb already formed by enzymatically reducing it back to Hb.

The amount of salt to add is based on the nitrite concentration in the water. A dissolved chloride-to-nitrite ratio of 16:1 completely suppresses M-Hb formation in channel catfish (Tomasso et al. 1979). However, a chloride-to-nitrite ratio of only 3:1 will prevent blood M-Hb from exceeding 50% and will usually also prevent further mortalities in otherwise healthy fish (Bowser et al. 1983). Although 50% M-Hb is not ideal, it is usually adequate to prevent mortalities and allows time for the pond water chemistry to correct itself (Bowser 1984). The currently recommended practice in commercial catfish culture is to monitor production ponds for nitrite at 3-day intervals during the months of the year (October–April) when brown blood disease is most common. If the nitrite concentration exceeds 0.5 mg/L, the chloride concentration in the pond water is also measured. If necessary, salt (sodium chloride) or calcium chloride is then applied to the pond to bring the ratio of chloride to nitrite up to 3:1. As a rule of thumb, adding 4.5 lbs salt (sodium chloride) per acre-foot of water yields a chloride concentration of 1 mg/L (Jensen 1989).

Individual valuable fish (e.g., brood stock) may also be treated for methemoglobinemia by injecting methylene blue intraperitoneally at a concentration of 10 mg/kg or by adding it to the water at 0.1–1.0 mg/L. Ascorbic acid (vitamin C) injected intravenously at 5 mg per day has also been used to reduce the amount of M-Hb in the blood (Wedemeyer and Yasutake 1978).

Visceral Granuloma and Nephrocalcinosis

Visceral granuloma and nephrocalcinosis are chronic degenerative inflammatory diseases primarily affecting the stomach and kidney of rainbow trout. Both diseases have long been known to be associated with environmental factors: diet in the case of visceral granuloma, and water chemistry in the case of nephrocalcinosis (Herman 1970; Landolt 1975). Because of their gradual onset, these diseases may go unnoticed until the fish are stressed by crowding or transportation at which time massive mortalities may occur.

Nephrocalcinosis is due to the formation of calcareous deposits in the kidney tubules. The deposits (similar to kidney stones in animals) are composed of insoluble precipitates containing calcium, phosphate, fluoide, and oxides of magnesium that resemble the minerals apatite and brucite. Rainbow trout reared in water containing 12–55 ppm CO_2 for up to 9 months developed nephrocalcinosis, the severity of which was proportional to the CO_2 concentration (Smart et al. 1979). At 55 ppm, clinical signs developed relatively quickly and growth was also seriously impaired. Nephrocalcinosis can also occur at lower dissolved CO_2 concentrations if the DO is supersaturated because of the hypercapnia that develops under hyperoxic rearing conditions (Schlotfeldt 1981).

Nephrocalcinosis has been observed in channel catfish and in marine fishes such as striped marlin and the striped mullet from the Salton Sea, California. In the case of marine fish, the calcium magnesium phosphate composition is similar to human kidney stones.

Water supplies high in bicarbonate hardness may facilitate kidney stone formation in trout especially if a diet high in protein is also fed. However, the dissolved carbon dioxide concentration and, to a lesser extent, the mineral composition of the diet are the primary environmental factors associated. For salmonids, the CO_2 recommendation to prevent nephrocalcinosis is 20 mg/L or less.

Blue Sac Disease (Hydrocele embronalis)

On occasion, yolk sac fry of salmonids will be observed to have an abnormal accumulation of bluish colored fluid in the yolk sac. White spots, due to coagulated (denatured) yolk proteins are frequently seen in the yolk sac itself accompanied by hemorrhaging from the vitelline blood vessels. Petechial hemorrhaging may also occur around the head and thoracic areas of the developing embryo. Due to excess pressure, exophthalmus (popeye) is common and growth and development are retarded.

Blue sac disease can be induced by excessive ammonia levels in the hatchery water supply or by a high level of metabolic wastes in incubation waters. All of the etiological factors involved in the development of this disease are not known. Indeed, as in the case of bacterial gill disease (*Flavobacterium branchiophilum*) the clinical signs might better be regarded as indicators that there is an underlying physiological disorder.

Blue sac disease can be prevented by water chemistry control so that accumulations of nitrogenous wastes do not occur. Conditions that restrict water flow around and through the eggs, such as crowding, fungal buildup, or clogged screens, should be corrected. Screens clogged with organic matter are especially bad because the water flow is reduced and nitrogenous materials are simultaneously released into the egg environment due to decomposition.

White-Spot (Coagulated-Yolk) Disease

The clinical signs of this noninfectious disease are multiple spots of white precipitated protein in the yolk material of (salmonid) eggs and fry. As with blue sac disease, the formation of coagulated protein is a clinical sign of an underlying physiological disorder termed *white-spot disease* (eggs) or *coagulated-yolk disease* (fry). It is well recognized that both diseases can result from a number of deleterious environmental factors including unfavorable temperatures, mineral deficient egg incubation water, rough handling resulting in physical injury to the perivitelline membrane, air supersaturation above the 102–103% level, heavy metals such as copper and zinc in the water supply, too much or too little water flow. Formerly, treating eggs for fungus with malachite green formulations containing more than about 0.08% zinc was also an occasional cause of white-spot disease.

The outcome of white-spot or coagulated-yolk disease is usually unfavorable. Some fry occasionally survive, but more often yolk absorption is never completed. The fins, especially the pectoral, turn white and the margins erode because of impaired blood circulation. Eventually most of these fry also die. A small amount of unabsorbed coagulated yolk often remains in the abdominal cavity.

The best defense against white-spot/coagulated-yolk disease is prevention. Rough handling, unfavorable temperatures and temperature fluctuations, and gas supersaturation above 102 to 103% should be avoided. Although malachite green is a known teratogen, it is still sometimes used for fungus control on incubating salmonid eggs. Formulations containing more than 0.08% zinc should be avoided. If the egg incubation water is very soft, adding $CaCl_2$ to raise the hardness to about 50 ppm ($CaCO_3$ equivalent) has been shown to greatly reduce mortalities (MacKinnon 1969). Adding NaCl, KCl or other sources of chloride ion using constant flow siphon bottles or metering pumps may also help.

Soft-Shell Disease (Soft-Egg Disease)

As the name implies, the principal clinical sign of this disease is a soft, flaccid chorion (shell) together with a very fragile perivitelline membrane that ruptures if the eggs are handled. Soft-shell disease apparently results from microscopic holes in the chorion that allow fluid to escape into the ambient water. The eggs become sticky, tend to clump together, and become very susceptible to fungal infections. The disease can be particularly troublesome at certain hatcheries whereas at others it is rarely noted.

Environmental factors have been implicated as well as infective agents. Disinfection with acriflavine, gentian violet (1 ppm), or other bactericidal agent has been successful in some cases, implying that an infective agent is involved. However, increasing the mineral content of the water supply can also be effective. For example, magnasite ore ($MgCO_3$) added to the headboxes of one soft water hatchery eliminated soft-shell problems at one salmonid hatchery. The water flow over the pieces of ore dissolved small amounts of Mg^{+2} and CO_3^{-2} and this effectively prevented the soft-shell condition from developing. Although more expensive, adding supplemental Ca^{++} to increase the final concentration to >10 mg/L was also beneficial (Wedemeyer et al. 1976).

Algal Toxins

The interactions of fish with the algae, a diverse group of aquatic plants ranging in size from microscopic planktonic cells to anchored seaweed fronds hundreds of feet long, are extremely important to the health and physiological quality of fish produced in either freshwater hatcheries, mariculture, or in natural systems. The algae most directly involved are mostly planktonic or filamentous and include the green algae (*Chlorophyta*), blue-green algae (*Cyanophyta*), and the golden-brown algae and diatoms (*Chrysophyta*). These and other algae both generate needed dissolved oxygen and serve, either directly or indirectly, as a source of food. Milkfish culture, for example, relies heavily on the maintenance of an ample supply of filamentous blue-green algae and diatoms in the ponds. The success of tilapia, carp, and mullet pond culture is also directly or indirectly dependent on algae. Juvenile salmonids released from conservation hatcheries depend for food on aquatic invertebrates and smaller fish that are supported by algae populations.

Unfortunately, algae can also harm fish as well as benefit them. Certain species of freshwater and marine algae produce potent nerve and liver toxins that periodically kill both hatchery and wild fish on a worldwide basis. For unknown reasons, blooms of toxic algae seem to be increasing in frequency and diversity in natural waters worldwide. Toxic algae growths can also accumulate on the concrete raceways commonly used in intensive fish culture. These growths may cause no problems unless normal pond cleaning breaks them up and releases the toxins in the water. Fish mortalities can then be substantial. Reasons for the production of nerve and liver toxins by algae are speculative, but may involve indirect protection against predation by zooplankton.

Nontoxic algae can also cause considerable harm. Heavy water blooms often terminate in a total oxygen depletion, killing large numbers of fish by asphyxiation. If fish are restricted to limited areas such as shallow water where algae have become greatly concentrated, the gill apparatus can become physically obstructed, again causing death by suffocation. In warmwater fish culture, an increased incidence of gill *Branchiomyces* (fungus) infections tends to occur during water blooms—especially of the blue-green algae. Finally, algae growth in warmwater fish ponds can cause production losses because of the production of off flavors and odors that are offensive enough to prevent commercial sales.

One of the most common problems caused by nontoxic algae is water blooms that terminate in an oxygen depletion. Such blooms usually occur following pond fertilization, but they may also develop as a result of nutrient enrichment from ordinary feeding practices or surface water runoff from agricultural or grazing lands. The first visible sign is usually a dense population of phytoplankton, often blue-green algae, developing in the water. Thick layers or scums soon form at the pond surface where there is more sunlight and the water is warmer. As the surface layer becomes thicker, light penetration into the deeper water is progressively reduced. The shading greatly reduces photosynthesis and the DO in the pond begins to decline except near the surface. Here, the water may actually become supersaturated with oxygen because of the intense photosynthesis. In the deeper water, benthic animals and plants begin to die and decompose providing nutrients that encourage the growth of aquatic bacteria. The increased bacterial metabolism further reduces the DO and adds undesirable amounts of carbon dioxide to the water as well. As anaerobic conditions develop in more and more of the pond, the oxygenated zone eventually becomes so small that fish cannot survive through the night. At this point, both fish and algae begin to die in large numbers. As they decompose, hydrogen sulfide, ammonia, and other toxic compounds are produced that act to accelerate the fish mortality rate. Eventually, the entire pond may become septic.

The best control method is prevention, but herbicides such as simazine, diquat, chelated copper, or copper sulfate are all effective in eliminating algae blooms. If herbicides are used, it is important to kill only part of the bloom at one time. If all of the algae are killed at once, the decreased photosynthesis and decomposing cells will only add to the oxygen depletion and accelerate its spread through the pond.

Fish kills caused by toxic algae follow a somewhat different pattern and can be difficult to diagnose. Water blooms of these algae may also terminate in an oxygen depletion, which then may be blamed for the fish mortalities while the toxic conditions go unnoticed. Fortunately, the differences between fish kills caused by toxic and nontoxic algae are significant. It is the recognition of these differences that provides the information needed for corrective action.

Toxins produced by the blue-green algae probably have the most potential to cause problems in freshwater fish culture. The blue-green algae are actually photosynthetic bacteria (cyanobacteria) rather than plants but the original terminology has persisted and is often used interchangeably. Most cyanobacteria are harmless but a dozen or so species from the genera *Anabaena*, *Microcystis*, and *Nodularia* produce potent nerve and liver toxins. Two of the most lethal cyanobacteria, *Anabaena flos-aquae* and *Microcystis aeureginosa* produce toxins that are not only lethal to fish, but can kill waterfowl and even large domestic animals. The latter usually happens when currents or winds have caused a water bloom of cyanobacteria to accumulate along a shore line. Thirsty animals may be undeterred by the off odors and taste and can easily drink enough water to ingest a fatal dose of the toxin-containing cells.

In warmwater fish ponds, the conditions favorable to a toxic cyanobacteria bloom include a water temperature of 15–30 C, pH 6–9, quiet or mild winds, and enrichment with nitrogen and phosphorous—either from feeding practices, fertilizer applications, or nutrients transported into the ponds by surface water runoff. Blooms of the true algae can also occur, but in nutrient-rich water, cyanobacteria blooms are more likely. As with the nontoxic algae, a dense blue-green growth forming in the pond is usually the first visible sign. A layer of cyanobacteria (commonly termed *pond scum*) typically accumulates at the surface where the most light is available. In hot weather, the scum layer can eventually become several inches thick. When the fish are fed, they may ingest the toxic cyanobacteria cells along with their food and be killed. Smaller fish are usually affected first because they normally tend to feed on small food particles at the surface. If floating diets are used, the problem is made worse and the entire fish population may eventually be killed. When the cyanobacteria cells die naturally or are killed by the herbicides commonly applied to stop the blooms, their toxins are released into the water. However, a fish kill from waterborne toxins is less likely than if whole cells are consumed because of the dilution that occurs.

The cyanobacteria produce two general types of toxins that can kill fish: neurotoxins and hepatotoxins. The neurotoxins are alkaloid compounds (see Adventitious Toxins in Chapter 4) that kill fish by disrupting normal communications between nerve and muscle. The hepatotoxins are termed *microscystins* and *nodularians* because they were first isolated from members of the genera *Microcystis* and *Nodularia*. These toxins consist of peptides (short chains of amino acids) arranged in ring structures. Hepatotoxins kill fish and other animals by disrupting the normal anatomical structure of the liver hepatocytes and causing massive internal bleeding.

Four of the neurotoxins have been characterized in detail. Two, anatoxin-a and anatoxin-a(s), are unique to the cyanobacteria. The others, saxitoxin and neosaxitoxin, are also produced by toxic marine algae from the golden-brown group responsible for red tide events and human paralytic shellfish poisoning.

Anatoxin-a stimulates muscle tissue to contract in a manner similar to acetylcholine, the natural neurotransmitter produced by motor neurons. Unfortunately, the enzyme acetylcholinesterase cannot immediately degrade anatoxin-a as it does acetylcholine and the muscles are continuously stimulated to contract. The clinical signs include twitching; erratic, convulsive swimming; (spastic) paralysis; and lethargy prior to death. After the muscles involved in respiration become paralyzed, the fish die from asphyxiation (suffocation).

Anatoxin-a(s) kills fish by inhibiting the enzyme acetylcholinesterase—the same physiological mechanism as death from exposure to the organophosphate insecticides (e.g., parathion). Again, muscle cells are overstimulated—this time because acetylcholine naturally produced at the myoneural junction is not degraded normally. The clinical signs are the same: twitching; erratic, convulsive swimming; lethargy; and death from spastic paralysis of the respiratory muscles.

Saxitoxin and neosaxitoxin also kill by disrupting normal communications between nerve and muscle. In this case, however, it is by preventing the motor neurons from releasing acetylcholine. The muscles are then unable to contract resulting in (flaccid) paralysis, lethargy, and the same eventual death by asphyxiation. As mentioned, saxitoxin and neosaxitoxin are also produced by the marine dinoflagellate algae responsible for the red tide events that cause great economic losses to the commercial shellfish industry worldwide.

No antidotes to cyanobacterial toxins are available. The only practical control methods are prevention, by managing nutrient loadings in the ponds to prevent water blooms, or to recognize that a toxic algae bloom is developing and destroy it with herbicides. Simazine, diquat, chelated copper, or copper sulfate are all effective in eliminating water blooms of toxic cyanobacteria. As with the nontoxic algae, however, it is important to kill only part of the bloom at one time. If all of the algae are killed at once, the decreased photosynthesis and decomposing cells can easily create an oxygen depletion and accelerate the fish mortality rate. In some cases, it may be practical to move fish to another pond until the water bloom dies out naturally.

Recognizing that the cause of death during cyanobacteria blooms is due to toxicosis rather than to an oxygen depletion is critical to taking corrective action. One indicator of algal toxins is a water bloom accompanied by fish kills with losses that peak during bright sunlight while there is still adequate DO in the pond. Also, fish suffering from algal toxicosis typically exhibit erratic, convulsive swimming behavior and then become lethargic before they die. Unfortunately, both of these clinical signs occur after-the-fact. A useful predictive indicator is that marked changes in the pond biota often occur as toxic cyanobacteria blooms develop. Usually a single species of blue-green algae becomes dominant and there may be a complete absence of zooplankton and benthic invertebrates. More detailed information on the taxonomy and ecology of the toxic blue-green algae can be found in the excellent review by Carmichael and Falconer (1993).

The golden-brown algae also produce toxins capable of killing fish. Again, most golden-brown algae are harmless but some of the marine species from the genera *Gymnodinium*, *Gyrodinium*, and *Alexandrium* (formerly *Gonyaulax*) produce an assemblage of potent neurotoxins (saxitoxin and related compounds) that can kill fish in microgram quantities. In freshwater, toxins produced by the golden-brown algae *Prymnesium parvum* pose a continuing threat to carp and tilapia culture. This algae can cause problems in brackish and estuarine fish culture as well.

Prymnesium parvum produces a toxin (prymnesin) that is released into the water where it enters fish through the gills. Cell membrane permeability is altered disrupting ionoregulation and nerve–muscle communication. Clinical signs include loss of equilibrium, gasping at the surface (due to breathing difficulties), and eventual death from asphyxiation. Fortunately, *Prymnesium* blooms are easily controlled by herbicides or by adding ammonium sulfate to the ponds at about 10 ppm. This salt concentration osmotically lyses the algal cells without harming other aquatic life. Copper sulfate at 2–3 ppm is also effective in controlling *Prymnesium*. However, to avoid fish toxicity problems, copper sulfate should never be used if the alkalinity is less than about 100 mg/L (as $CaCO_3$).

The marine species of *Gonyaulax* and the other toxic golden-brown algae are reddish-colored motile unicellular organisms (dinoflagellates) that produce an assemblage of potent neurotoxins including saxitoxin. Blooms of these dinoflagellates occurring in nearshore areas can be dense enough to color the water resulting in so-called red tides. Shellfish feeding on these algae accumulate the neurotoxins but are usually unaffected by them. Humans who consume the shellfish can become seriously ill or die from a syndrome termed *paralytic shellfish poisoning* (PSP). As discussed, saxitoxin and related PSP toxins effectively prevent the propagation of electrical impulses. The central nervous system is disrupted and neuromuscular transmission by motor neurons is blocked. The end result is paralysis and death from respiratory collapse. The lethal dose for fish is unclear, but the human lethal dose is known to be <500 μg. The recurring presence of PSP toxins in shellfish resulting from water blooms of these toxic algae is a continuing public health problem worldwide and causes serious economic losses to the shellfish aquaculture industry. Red tides due to toxic dinoflagellate blooms seem to be increasing in frequency and have become a worldwide phenomenon in both temperate and tropical marine waters.

Although the major economic impact of PSP producing dinoflagellates is probably on shellfish aquaculture, their toxins also cause periodic mortalities in marine fish populations, as well as adding a certain degree of risk to marine net pen culture. Copepods feeding on *Gonyaulax* and other dinoflagellates accumulate the PSP toxins and become vectors to fish. Similarly, if sufficient numbers of toxic salmon, herring, or other fish are consumed by predators, the predators may then be killed. Finfish have been implicated as vectors for PSP toxins in the deaths of both seabirds and marine mammals. Kills of marine fish such as herring, cod, and menhaden have all occurred. Anadromous salmonids such as chinook and coho salmon and steelhead trout that spend their first year rearing in freshwater would not encounter PSP toxins as very young fish. However, these species may be vulnerable after they migrate into the ocean if their prey items contain PSP toxins. Species such as juvenile pink salmon and herring that spend their first spring and summer rearing in estuaries and nearshore marine areas are vulnerable to PSP toxins in their food items (Erickson 1988). Although the blooms of *Gonyaulax* and other algal species that produce PSP toxins now occur nearly worldwide, it seems unlikely that they have any significant effect on the stock size of oceanic fish populations.

PSP toxins do add a certain degree of risk to marine aquaculture operations. Rainbow trout, Atlantic salmon, and yellowtail in net pen culture have all been killed by *Gonyaulax* blooms. Clinical signs include disorientation; erratic, convulsive swimming; and lethargy prior to death (Saito et al. 1985). The ingredients for a *Gonyaulax* kill of fish in net pen culture are: the water bloom itself; zooplankton (copepods) that can consume *Gonyaulax* and become toxic; and species of fish in the net pens that will feed on the toxic zooplankton. The present intensification and spread of toxic dinoflagellate blooms throughout the world suggests that the effects of PSP toxins will become increasingly important to the health and physiological condition of fish produced in marine aquaculture systems.

Two other marine algae, the motile chloromonad flagellate *Heterosigma* spp. and the diatom *Chaetoceros* spp. have now developed into a threat to the health of fish in intensive net pen culture, particularly of Pacific and Atlantic salmon. Caged fish exposed to *Chaetoceros* blooms at a concentration of >5 cells/ml seawater suffer gill lacerations, hemorrhaging, inflammation, and secondary infections when the spines of these diatoms become embedded in the secondary lamellae. The copious mucus discharged by the gills in response to the embedded spines also hinders normal gas exchange. Clinical signs displayed by fish affected by *Chaetoceros* include gasping at the surface and a characteristic periodic coughing response. Increased mortality from bacterial kidney disease and vibriosis have been reported in chinook and coho salmon held in seawater net pens during *Chaetoceros* blooms at the otherwise sublethal concentration of <5 cells/ml (Albright et al. 1993).

The physiological mechanism by which water blooms of *Heterosigma* kill fish is unknown, but serious losses of salmon in marine net pen culture caused by this marine flagellate are becoming increasingly common worldwide. One species, *Heterosigma akashiwo*, has been linked to kills of caged fish in the coastal waters of Japan, North America, and Scotland among other places. Blooms are associated with water temperatures of 15°C or more and can last for several months. Runoff from rivers contributes to the initiation of a bloom by stratifying the seawater column and adding micronutrients. Factors contributing to the long persistence of *Heterosigma* blooms include its tolerance to wide salinity changes, which enables this flagellate to exploit the nutricline by vertical migration, and lack of grazing by zooplankton or other predators in the stratified layers (Taylor and Haigh 1993). Chemical control agents have not been successful. Other control measures that have been partly successful include shading the net pens to induce the flagellates to move to lighted areas, skirting the net pens, or lowering the pens deep enough to allow the fish to move below the bloom.

The marine dinoflagellate *Gambierdiscus toxicus* produces a toxin (ciguatera toxin) that occurs at times in food fishes from the Caribbean Islands and other tropical waters. *Gambierdiscus toxicus* lives on the surface of macroalgae (seaweed) found in coral reef areas. Small fish feeding on this epiphytic algae accumulate the toxin. Larger fish preying on the smaller species are not killed but their flesh becomes toxic and unsafe for human consumption.

A comparison of significant differences and similarities that can be helpful in determining whether fish kills might have been due to an oxygen depletion, a toxic algal bloom, or exposure to other toxins such as pesticides is given in Table 3.7. Details of the taxonomy of harmful marine algae and the environmental conditions necessary for bloom development can be found in Gaines and Taylor (1986).

TABLE 3.7. Characteristics of fish kills caused by toxic algae, oxygen depletions, and contaminants such as pesticides.

Physical Signs Associated with the Mortality	Cause of Mortality		
	Toxic Algae	Oxygen Depletion	Pesticide Toxicity
Fish behavior	Convulsive, erratic swimming; lethargy	Swimming at surface, "gasping"	Convulsive, erratic swimming
Species selectivity	None, all species affected	None if depletion is total, carp may survive partial depletion	Some species killed before others depending on sensitivity and pesticide
Fish size	Small fish killed first, eventually all sizes killed	Large fish killed first	Small fish killed first
Time of fish kill	Only during hours of bright sunlight	Night and early morning	Any time, day or night
Plankton abundance	One algal species predominates, little zooplankton present	Algae dying, little zooplankton present	Insecticide, zooplankton absent, algae abundant; herbicide, algae may be absent
DO levels	DO very high, may be supersaturated near surface	Less than 2 ppm, usually below 1 ppm	Normal range
Water pH	9.5 and above	6.0–7.5	7.5–9.0
Water color and odor	Dark green, brown, golden color; musty odor may be present	Brown, gray, or black	Normal color, little or no off odor
Algae bloom present	Algae abundant, one species predominates	Many algal cells dead or dying	Normal mixed species; if herbicide, algae may be reduced or absent

Source: information compiled from Wedemeyer et al. (1976).

REFERENCES

Ahne, W., and K. Wolf. 1980. Viruserkrankungen der Fische. Pages 123–156 *in* H. H. Reichenbach-Klinke (ed.), Krankheiten und Schädigungen der Fische. 2nd ed., Fischer Verlag, Stuttgart, FRG.

Alabaster, J. S., and R. Lloyd. 1980. Water Quality for Freshwater Fish. Butterworth, Sydney.

Albright, L. J., C. Z. Yang, and S. Johnson. 1993. Sub-lethal concentrations of the harmful diatoms, *Chaetoceros concavicornis* and *C. convolutus,* increase mortality rates of penned Pacific salmon. Aquaculture 117:215–225.

Alderdice, D. F., and J. O. T. Jensen. 1985. Assessment of the influence of gas supersaturation on salmonids in the Nechako River in relation to Kemano Completion. Canada Technical Report Fisheries and Aquatic Science No. 1386.

Allanson, B. R., A. Bok, and N. I. Van Wyk. 1971. The influence of exposure to low temperature on *Tilapia mossambica* Peters (Cichlidae). II. Changes in serum osmolarity, sodium, and chloride ion concentrations. Journal of Fish Biology 3:181–185.

Basu, S. P. 1959. Active respiration of fish in relation to ambient concentrations of oxygen and carbon dioxide. Journal of the Fisheries Research Board of Canada 16:175–212.

Beckman, B. R., and W. S. Zaugg. 1988. Copper intoxication in chinook salmon (*Oncorhynchus tshawytscha*) induced by natural spring water: effects on gill Na^{+}-K^{+} ATPase, hematocrit, and glucose. Canadian Journal of Fisheries and Aquatic Sciences 45:1430–1435.

Bowser, P. R. 1984. Brown blood disease (Methemoglobinemia) of fishes. Fish Disease Leaflet 70, U.S. Department of the Interior, Fish and Wildlife Service, Washington, D.C.

Bowser, P. R., W. W. Falls, J. Van Zandt, N. Collier, and J. D. Phillips. 1983. Methemoglobinemia in channel catfish: methods of prevention. Progressive Fish-Culturist 45:154–158.

Bradley, T. M., and R. W. Rourke. 1985. The influences of addition of minerals to rearing water and smoltification on selected blood parameters of juvenile steelhead trout, *Salmo gairdneri* Richardson. Physiological Zoology 58:312–319.

Carmichael, W. W., and I. R. Falconer. 1993. Algal toxins in seafood and drinking water. Pages 45–67 *in* I. R. Falconer (ed.), Diseases Related to Freshwater Blue-Green Algal Toxins, and Control Measures. Academic Press, New York.

Cleveland, L., E. E. Little, R. H. Wiedmeyer, and D. B. Buckler. 1989. Chronic no-observed-effect concentrations of aluminum for brook trout exposed in low-calcium dilute acidic water. Pages 229–245 *in* T. E. Lewis (ed.), Environmental Chemistry and Toxicity of Aluminum. Lewis Publishers, Chelsea, Michigan.

Colt, J. 1984. Seasonal changes in dissolved-gas supersaturation in the Sacramento River and possible effects on striped bass. Transactions of the American Fisheries Society 113:655–665.

Colt, J. 1989. Dissolved gas supersaturation. Pages 2–9 *in* A. Greenberg, R. Trussel, L. Clesceri (eds.), Standard Methods for the Examination of Water and Waste Water, 16th ed. American Public Health Association, Washington, D.C.

Colt, J. E., and G. Tchobanoglous. 1981. Design of aeration systems for aquaculture. Pages

138–148 *in* L. Allen and E. Kinney (eds.), Bioengineering Symposium for Fish Culture. American Fisheries Society, Bethesda, Maryland.

Colt, J., and D. A. Armstrong. 1981. Nitrogen toxicity to crustaceans, fish, and molluscs. Pages 34–47 *in* L. J. Allen and E. C. Kinney, (eds.), Bioengineering Symposium for Fish Culture. American Fisheries Society, Bethesda, Maryland.

Colt, J. K., K. Orwicz, and G. R. Bouck. 1991. Water quality considerations and criteria for high-density fish culture with supplemental oxygen. American Fisheries Society Symposium 10:372–385.

Das, M. K., and R. K. Das. 1993. A review of the fish disease epizootic ulcerative syndrome in India. Environmental Ecology 11:134–145.

Davis, J. C., and I. G. Shand. 1978. Acute and sublethal copper sensitivity, growth and saltwater survival in young Babine Lake sockeye salmon. Canadian Fisheries and Marine Service Technical Report 847. Department of Fisheries and Oceans, Ottawa, Canada.

Dawson, V. K. 1986. Computer program calculation of gas supersaturation in water. Progressive Fish-Culturist 48:142–146.

Edsall, D., and C. E. Smith. 1989. Performance of rainbow trout reared in oxygen supersaturated water. Salmonid 13:15–17.

EPA (Environmental Protection Agency). 1986. Quality Criteria for Water. EPA 440/5–86–001, U.S. Environmental Protection Agency, Washington, D.C.

Erickson, G. M. 1988. The effect of *Gonyaulax castenella* toxins on chum and pink salmon smolts (*Oncorhynchus keta* and *O. gorbuscha*), coho salmon fry (*O. kisutch*), and Pacific herring juveniles (*Clupea harengus pallase*). M.S. thesis, University of Washington, Seattle, Washington.

Fryer, J. L., and C. N. Lannan. 1993. The history and current status of *Renibacterium salmoninarum*, the causative agent of bacterial kidney disease in Pacific salmon. Fisheries Research 17:15–33.

Fuss, J. T. 1986. Design and application of vacuum degassers. Progressive Fish-Culturist 48:215–221.

Gaines, G., and F. J. R. Taylor. 1986. A mariculturist's guide to potentially harmful marine phytoplankton of the Pacific coast of North America. Information Report No. 10, B.C. Ministry of Fisheries, Victoria, B.C., Canada.

Groberg, W. J., J. S. Rhovec, and J. L. Fryer. 1983. The effects of water temperature on infection and antibody formation induced by *Vibrio anguillarum* in juvenile coho salmon (*Oncorhynchus kisutch*). Journal of the World Mariculture Society 14:240–248.

Haswell, M. S., G. J. Thorpe, L. E. Harris, T. C. Mandis, and R. E. Rauch. 1982. Millimolar quantities of sodium salts used as prophylaxis during fish hauling. Progressive Fish-Culturist 44:179–182.

Herman, R. L. 1970. Visceral granuloma and nephrocalcinosis. Fish Disease Leaflet 32, U.S. Fish and Wildlife Service, Washington, D.C.

Houston, A. H., and T. F. Koss. 1984. Erythrocytic haemoglobin, magnesium and nucleoside triphosphate levels in rainbow trout exposed to progressive heat stress. Journal of Thermal Biology 9:159–164.

Huner, J. V., and H. J. Dupree. 1984. Pond management. Pages 17–43 *in* H. Dupree and J. Hunter (eds.), Third Report to the Fish Farmers: The Status of Warmwater Fish Farming and Progress in Fish Farming Research. U.S. Fish and Wildlife Service, Washington, D.C.

Jensen, G. 1989. Handbook for common calculations in finfish aquaculture. Publication 8903 Louisiana State University, Agricultural Center. Baton Rouge, Louisiana.

Kils, U. 1977. The salinity effect on aeration in mariculture. Meeresforsch 25:755–759.

Kincheloe, J. W., G. A. Wedemeyer, and D. L. Koch. 1979. Tolerance of developing salmonid eggs and fry to nitrate exposure. Bulletin of Environmental Contamination and Toxicology 23:575–578.

Klontz, G. W. 1993. Environmental requirements and environmental diseases of salmonids. Pages 333–342 *in* M. Stoskopf (ed.), Fish Medicine. W. B. Saunders, Phiadelphia, Pennsylvania.

Knittel, M. D., G. A. Chapman, and R. R. Garton. 1980. Effects of hydrostatic pressure on steelhead survival in air supersaturated water. Transactions of the American Fisheries Society 109:755–759.

Krise, W. F., and J. W. Meade. 1988. Effects of low level gas supersaturation on lake trout (*Salvelinus namaychus*). Canadian Journal of Fisheries and Aquatic Science 45: 666–674.

Lall, S. P., W. D. Paterson, J. A. Hines, and N. J. Hines. 1985. Control of bacterial kidney disease in Atlantic salmon *Salmo salar* L., by dietary modification. Journal of Fish Diseases 8:113–124.

Landolt, M. L. 1975. Visceral granuloma and nephrocalcinosis in trout. Pages 793–801 *in* L. Ribelin and G. Miyake (eds.), Pathology of Fishes. University of Wisconsin Press, Madison, Wisconsin.

Lorz, H., and B. P. McPherson. 1976. Effects of copper or zinc in fresh water on the adaptation to sea water and ATPase activity, and the effects of copper on migratory disposition of coho salmon (*Oncorhynchus kisutch*). Journal of the Fisheries Research Board of Canada 33:2023–2030.

Lorz, H., S. Glenn, R. Williams, C. Kunkel, L. Norris, and B. Loper. 1978. Effect of selected herbicides on smolting of coho salmon. U.S. Environmental Protection Agency, Grant Report R-804283. Oregon Department of Fish and Wildlife, Corvallis, Oregon.

Mackinnon, D. F. 1969. Effect of mineral enrichment on the incidence of white-spot disease. Progressive Fish-Culturist 31:74–78.

Marking, L. L. 1987. Gas supersaturation in fisheries: causes, concerns, and cures. Fish and Wildlife Leaflet 9, United States Department of the Interior, Fish and Wildlife Service, Washington, D.C.

Marking, L. L., and T. D. Bills. 1982. Factors affecting the efficiency of clinoptilolite for removing ammonia from water. Progressive Fish-Culturist 44:187–189.

Marking, L. L., Dawson, V. K., and F. R. Crowther. 1983. Comparison of column aerators and a vacuum degasser for treating supersaturated culture water. Progressive Fish-Culturist 45:81–83.

Mathias, J. A., and J. Barcia. 1985. Gas supersaturation as a cause of early spring mortality of stocked trout. Canadian Journal of Fisheries and Aquatic Science 42:268–279.

McDonough, P. M., and E. A. Hemmingsen. 1985. Swimming movements initiate bubble formation in fish decompressed from elevated gas pressures. Comparative Biochemistry and Physiology 81A:209–212.

Meade, J. W. 1985. Allowable ammonia for fish culture. Progressive Fish-Culturist 47:135–145.

Messer, J. J., J. Ho, and W. J. Grenney. 1984. Ionic strength correction for extent of ammonia ionization in freshwater. Canadian Journal of Fisheries and Aquatic Science 41:811–815.

Nickum, J. D. 1966. Some effects of sudden temperature changes upon selected species of freshwater fishes. Ph.D. dissertation. Southern Illinois University, Carbondale, Illinois.

Owsley, D. E. 1981. Nitrogen gas removal using packed columns. Pages 71–82 *in* L. J. Allen and E. C. Kinney, (eds.), Proceedings of the Bioengineering Symposium for Fish Culture, American Fisheries Society, Bethesda, Maryland.

Post, G. B. 1987. Textbook of Fish Diseases. TFH Publications, Neptune, New Jersey.

Redding, J. M., and C. B. Schreck. 1987. Physiological effects on coho salmon and steelhead of exposure to suspended solids. Transactions of the American Fisheries Society 116:737–744.

Reichenbach-Klinke, H. H. 1981. The influence of temperature and temperature changes on the outbreak and intensity of fish diseases. Pages 103–107 *in* Proceedings of the World Symposium on Aquaculture in Heated Effluents and Recirculation Systems, vol. II. Heenemann Verlagsgesellschaft, Berlin.

Russo, R. C., and R. V. Thurston. 1977. The acute toxicity of nitrite to fishes. Pages 118–131 *in* R. A. Tubb (ed.), Recent Developments in Fish Toxicology. National Environmental Research Service, Ecological Research Series EPA–600/3–77–085. U.S. Environmental Protection Agency, Washington, D.C.

Russo, R. C., R. V. Thurston, and K. Emerson. 1981. Acute toxicity of nitrite to rainbow trout (*Salmo gairdneri*): effects of pH, nitrite species, and anion species. Canadian Journal of Fisheries and Aquatic Science 38:389–393.

Saito, T., T. Noguchi, T. Takeuchi, S. Kanimura, and K. Hashimoto. 1985. Ichthyotoxicity of paralytic shellfish poison. Bulletin of the Japanese Society of Scientific Fisheries 51:257–260.

Saunders, R. L., E. B. Henderson, P. R. Harmon, C. E. Johnston, and J. G. Eales. 1983. Effects of low environmental pH on smolting of Atlantic salmon (*Salmo salar*). Canadian Journal of Fisheries and Aquatic Sciences 40:1203–1211.

Schlotfeldt, H. J. 1981. Some clinical findings of a several years survey of intensive aquaculture systems in Northern Germany with special emphasis on gill pathology and nephrocalcinosis. Pages 109–119 *in* K. Tiews (ed.), Proceedings of the World Symposium on Aquaculture in Heated Effluents and Recirculation Systems, vol. II. Heenemann Verlagsgesellschaft, Berlin.

Schwedler, T. E., and C. S. Tucker. 1983. Empirical relationship between percent methemoglobin in channel catfish and dissolved nitrite and chloride in ponds. Transactions of the American Fisheries Society 112:117–119.

Servizi, J. A., and D. W. Martens. 1978. Effects of selected heavy metals on early life of sockeye and pink salmon. International Pacific Salmon Fisheries Commission, Progress Report 39.

Smart, G. R., D. Knox, J. G. Harrison, R. H. Richards, and C. B. Cowey. 1979. Nephrocalcinosis in rainbow trout *Salmo gairdneri* Richardson: the effect of exposure to elevated CO_2 concentrations. Journal of Fish Diseases 2:279–289.

Soderberg, R. W. 1995. Flowing Water Fish Culture. Lewis Publishers, Boca Raton, Florida.

Stevens, D. G., A. V. Nebeker, and R. J. Baker. 1980. Avoidance responses of salmon and trout of air-supersaturated water. Transactions of the American Fisheries Society 109:751–754.

Sun, L. T., G. R. Chen, and C. F. Chang. 1992. The physiological responses of tilapia exposed to low temperatures. Journal of Thermal Biology 17:149–153.

Taylor, F. J. R., and R. Haigh. 1993. The ecology of fish-killing blooms of the chloromonad flagellate *Heterosigma* in the Strait of Georgia and adjacent waters. Pages 705–710 *in* T. Smayda and Y. Shimizu (eds.), Toxic-Phytoplankton Blooms in the Sea. Elsevier, Amsterdam.

Tomasso, J. R., B. A. Simco, and K. B. Davis. 1979. Chloride inhibition of nitrite induced methemoglobinemia in channel catfish (*Ictalurus punctatus*). Journal of the Fisheries Research Board of Canada 36:1141–1144.

Tomasso, J. R., M. I. Wright, B. A. Simco, and K. B. Davis. 1980. Inhibition of nitrite-induced toxicity in channel catfish by calcium chloride and sodium chloride. Progressive Fish-Culturist 42:144–146.

Tucker, C. S. 1993. Water analysis. Pages 166–197 *in* M. Stoskopf (ed.), Fish Medicine, W. B. Saunders, Philadelphia, Pennsylvania.

Tucker, C. S., and E. H. Robinson. 1990. Channel Catfish Farming Handbook. Van Nostrand Reinhold, New York.

Watten, B. J., J. Colt, and C. Boyd. 1991. Modeling the effect of dissolved nitrogen and carbon dioxide on the performance of pure oxygen absorption systems. American Fisheries Society Symposium 10:474–481.

Wedemeyer, G. A. 1973. Some physiological aspects of sublethal heat stress in the juvenile steelhead trout (*Salmo gairdneri*) and coho salmon (*Oncorhynchus kisutch*). Journal of the Fisheries Research Board of Canada 30:831–834.

Wedemeyer, G. A., and C. P. Goodyear. 1984. Diseases caused by environmental stressors. Pages 424–434 *in* O. Kinne (ed.), Diseases of Marine Animals, vol. 4, part 1: Introduction, Pisces. Biologische Anstalt Helgoland, Hamburg, FRG.

Wedemeyer, G. A., and W. T. Yasutake. 1978. Prevention and treatment of nitrite toxicity in juvenile steelhead trout (*Salmo gairdneri*). Journal of the Fisheries Research Board of Canada 35:822–827.

Wedemeyer, G. A., F. P. Meyer, and L. Smith. 1976. Environmental Stress and Fish Diseases. TFH Publications, Neptune, New Jersey.

Wright, P. A., and C. M. Wood. 1985. An analysis of branchial ammonia excretion in the freshwater rainbow trout: effects of environmental pH change and sodium uptake blockade. Journal of Experimental Biology 114:329–353.

Zaugg, W. S., and L. R. McLain. 1976. Influence of water temperature on gill sodium, potassium-stimulated ATPase activity in juvenile coho salmon (*Oncorhynchus kisutch*). Comparative Biochemistry and Physiology 54A:419–421.

Zaugg, W. S., and H. H. Wagner. 1973. Gill ATPase activity related to parr–smolt transformation and migration in steelhead trout (*Salmo gairdneri*): influence of photoperiod and temperature. Comparative Biochemistry and Physiology 45B:955–965.

4
Effects of Fish Cultural Procedures

INTRODUCTION

In addition to challenges from the water quality conditions discussed in Chapter 3, the physiological systems of fish in intensive culture must also cope with stressful hatchery procedures. Of these, the effects on health and physiological condition caused by crowding, handling, and transportation are probably the most widely appreciated by fish culturists. However, procedures such as disease treatments and feeding formulated diets are also important in this regard because of side effects of the drugs used and the adventitious toxins feeds can contain. Most of the pathophysiological problems associated with fish cultural procedures manifest themselves relatively quickly as increased mortality rates or reduced growth, however, some can be quite insidious. For example, the adverse effects of fish cultural procedures on the physiological condition of anadromous salmonids often become apparent only after the smolts attempt to enter seawater. Fortunately, many of the health and physiological quality problems associated with fish cultural procedures are now beginning to be understood. This understanding can serve as the basis for management actions that will minimize or prevent the costly production losses that often otherwise occur.

CROWDING

Of all the rearing conditions that can adversely affect the general health and physiological condition of fish in intensive culture, crowding—that is, approaching or exceeding the density tolerance limits of the species in question—is undoubtedly one of the most common. The term *crowding* is often loosely used to describe a high fish-loading density (calculated as the weight of fish/unit of water flow). However, (*sensu stricto*) crowding more correctly describes the behavioral requirements of the fish for physical space and is expressed in terms of the weight density or the weight of fish/unit volume of water. Although the two concepts are related, the fish loading density (weight per unit of flow) is actually a measure of the carrying capacity of the water.

Carrying capacity in the ecological sense is defined as the maximum number of individuals that the resources of a particular habitat can support. In fish hatchery management, carrying capacity refers to the water in a rearing unit and is usually expressed as a weight loading (fish weight/unit of water flow) rather than as numbers of fish. The carrying capacity of a given amount of water is determined by the oxygen consumption rate of the fish and by their tolerance to the ammonia, carbon dioxide, and other toxic metabolic wastes they produce. In warmwater pond culture and in water reuse systems in which nitrifying biofilters are used for ammonia removal, carrying capacity may be limited by nitrite accumulation as well. Tolerance to high fish loadings is thus related to, but different from, tolerance to crowding; that is, to the behavioral requirements of the particular fish for physical space (expressed as fish weight/unit water volume).

Both fish density (crowding) and fish loading (carrying capacity of the water) are highly significant biological criteria for intensive culture systems because economic considerations usually dictate that maximum use be made of both water and space. In intensive culture systems, the biomass that a given flow of water can support is usually first limited by metabolism because this determines the rate of oxygen consumption and waste production. Behavioral requirements for space are usually the second limiting factor. Although space is important, it is usually not available physical space (volume) that limits growth but the availability of the high quality water and flow rates needed to provide oxygen and dilute metabolic waste.

In raceways and tanks, the water exchange rate (R), which is a function of flow, is the parameter that actually limits carrying capacity because it controls the amount of oxygen that will be supplied and the rate at which ammonia and carbon dioxide are carried away. For example, a simple calculation shows that a water flow of 1 Lpm will supply about 600 mg O_2/h if the DO is 10 mg/L or 1.4 g of O_2 each 24 hours for each mg/L of dissolved oxygen that the water contains. In pond culture, oxygen is also supplied by diffusion through the surface and water inflow is not as important as surface area and pond volume in determining carrying capacity. The fish pathogen load of the water supply is also a very important factor affecting the carrying capacity of hatchery water supplies but it is difficult to quantify and rarely directly considered.

As a first approximation, flow rates (exchange rates) that will meet minimum physiological requirements can be calculated from the rate of oxygen consumption and sensitivity to ammonia exposure of the species in question. For salmonids, this is the water exchange rate that will maintain the DO concentration above 6 mg/L and the un-ionized ammonia below 0.02 mg/L at the outflow end of the rearing unit. The maximum flow loading can also be determined experimentally by a simple bioassay. The water inflow into a rearing unit containing a known weight of fish is progressively reduced until the effluent DO decreases to 5–6 ppm and the ammonia concentration rises to 0.02 mg/L (for salmonids). However, the cumulative oxygen consumption bioassay developed by Meade (1991), in which the oxygen removal that results in a predetermined reduction in growth, say 10%, gives better results and is currently the recommended method of determining production capacity. As is shown in Table 4.1, the oxygen consumption of rainbow trout is considerably higher than either chinook or coho salmon. Based on this, fewer trout can theoretically be reared in a given flow of water than equivalent-sized salmon. In practice, however, carrying capacities for trout are usually higher than for either Pacific or Atlantic salmon due to behavioral factors such as domestication and greater physiological tolerance to crowding stress.

An alternative method of expressing carrying capacity is in terms of the feeding rate per unit of water flow (lbs food fed/gpm). This method is usually termed the *feed carrying capacity*. That is, the desired water quality conditions can be achieved by adjusting the ration fed as well as by adjusting the fish loading. The factors that limit carrying capacity, oxygen consumption, ammonia excretion, and

TABLE 4.1. Oxygen consumption rates of rainbow trout and coho salmon (10°C) at swimming and feeding activity levels typical of hatchery raceway conditions.

Fish Size (grams)	Length (cm)		O_2 Consumption (mg/kg/h)	
	Trout	Salmon	Trout	Salmon
0.5	3.5	3.9	421.7	307.6
1.0	4.5	4.9	383.2	268.9
1.5	5.1	5.6	362.3	248.6
2.0	5.7	6.2	348.2	235.1
3.0	6.4	7.1	329.3	217.3
4.0	7.1	7.8	316.5	205.5
5.0	7.7	8.5	306.9	196.8
10.0	9.7	10.6	278.9	172.1
25.0	13.1	14.4	245.8	144.0
50.0	16.5	18.0	223.3	125.9
75.0	18.8	20.7	211.1	116.4
100.0	20.8	22.9	202.9	110.1

Source: Calculated from $O_2 = KT^nW^m$ (Liao 1971).

Note: A water flow of 1 Lpm will supply 600 mg O_2/h if the DO is 10 mg/L.

carbon dioxide production, are a predictable function of the amount of food consumed and it does not matter if a unit weight of feed is consumed by small or large fish. As an example, feeding 1 kg of a typical commercial diet to rainbow trout results in the consumption of 200 g of O_2, the production of 280 g of CO_2, and the excretion of 30 g of total ammonia (Meade 1991). Thus, limiting the feeding rate to 0.2 lbs dry diet per day per gpm water inflow will also maintain the DO above 5 mg/L and the total ammonia levels below 0.1 mg/L if the DO in the incoming water is 10 mg/L or more.

Two other measures of carrying capacity and crowding have been developed by Piper and co-workers that relate both fish weight and size to the water flow rate and space requirements of the species (Piper et al. 1982). These are termed the *flow index* (lbs fish/gpm/inch) and the *density index* (lbs fish/ft^3/inch) respectively. The flow index relates the carrying capacity of water to the amount of oxygen needed as fish size changes. The density index expresses tolerance to crowding (the behavioral need for space) as a function of fish size.

As mentioned, crowding should be measured by the tolerance of the fish to density (as determined by their requirement for physical space) rather than as fish loading (carrying capacity of the water for the species in question). The two concepts have been mathematically interrelated by Westers (1984) using the water exchange rate (R).

$$\text{Fish density (kg/m}^3) = \text{Fish loading (kg/Lpm)} \times (R/0.06 \text{ m}^3)$$
$$\text{Fish loading (kg/Lpm)} = \text{Fish density (kg/m}^3) \times (0.06 \text{ m}^3)/R$$

where:

R = water exchange rate per hour and 0.06 m3 = volume of a flow of one Lpm for one hour.

Thus, if there is one water exchange per hour ($R = 1$), a fish loading of 1.0 kg/Lpm results in a fish density of 16.7 kg/m^3. If the water flow rate is kept constant, increasing the density will also increase the fish loading. For salmonids, a water exchange rate of at least 2 ($R = 2$) is considered highly desirable for raceway rearing although it is sometimes unattainable.

The pathogen load of the water supply is as relevant to density tolerance as it is to carrying capacity but again has not been used directly because it is difficult to quantify.

Density tolerance has been an elusive concept to quantify. Densities that are too conservative waste space whereas densities that are too high cause stress, disease problems such as fin erosion, and can reduce subsequent seawater survival as well. The degree of crowding that fish can tolerate is species specific. For rainbow trout, the traditional consensus has been that they should be held at a density no greater than one-half their length in inches (or 3.4 times their length in cm). That is, at a density index of less than 0.5 (lbs/ft^3·in) or 3.4 (kg/m^3·cm). Densities of 80–

350/kg/m^3 have been successfully achieved in the intensive raceway culture of channel catfish (Ray 1981). However, density tolerance is usually a less important concern for most warmwater fishes because, other than the channel catfish, they are normally cultured extensively. Carrying capacities for the water used in intensive catfish culture are determined using the concepts derived for coldwater fishes.

Rearing densities that violate physiological and behavioral requirements for space (volume) will result in debilitated health and physiological condition, reduced food conversion, reduced growth, and ultimately higher mortality. Anadromous fish undergoing the parr–smolt transformation are more sensitive than parr or nonanadromous species. For example, coho salmon parr moved to densities over 16 kg/m^3 suffer mild blood chemistry disturbances such as hyperglycemia and hypochloremia. The same density causes severe blood chemistry disturbances and eventual mortality in smolting coho. In contrast, nonanadromous strains of rainbow trout show no adverse physiological effects until the density exceeds 26 kg/m^3 (Wedemeyer 1982).

It has long been known that nonanadromous salmonids can be reared at extreme densities such as the historic 540 kg/m^3 figure attained by Buss for rainbow trout in vertical silos (Buss 1981). Crowding levels (densities) that seem very high can be made to be satisfactory by using very high R values to meet physiological needs for respiration and metabolite dilution, or by rearing fish in net pens. In the latter case, R is usually low but the water volume is usually large enough to approximate infinite dilution of their metabolic wastes. Thus, rainbow trout held in net pens suspended in circular tanks provided with sufficient flow to maintain loadings at 800 g/Lpm or less could be reared at densities as high as 267 kg/m^3; equivalent to a density index of 11.1 g/L·cm^{-1} with no adverse effects on growth, condition factor or other clinical sign of physiological stress (Kebus et al. 1992). In production hatcheries of course, densities as high as these would not be practical. Traditionally, most conservation hatcheries have been designed to rear nonanadromous salmonids at the much lower densities of 30–50 kg/m^3 because lower water velocities (<3 cm/sec) that will allow solids to settle can be used. Biologically, however, experience has shown that densities of up to 120 kg/m^3 could easily be attained (Westers 1984). Lake trout have been reared at densities up to about 160 kg/m^3 (Soderberg 1995), catfish in raceways at 350 kg/m^3 or more (Ray 1981), and carp in pond culture as high as 278 kg/m^3 (Hickling 1971). In contrast, raceway rearing densities of 10–15 kg/m^3 may be more prudent for anadromous Pacific salmon in view of the evidence that the ocean survival of chinook smolts reared at raceway density levels of 30–40 kg/m^3 or more can be reduced (Ewing and Ewing 1995). A consensus is developing that other anadromous salmonids, particularly Atlantic salmon and coho salmon, should also be reared at low densities.

As mentioned, a water velocity of 3 cm/sec (0.09 ft/sec) is the minimum that will prevent solids such as feces and uneaten food from settling out and help keep a raceway clean (Youngs and Timmons 1991). Because water velocity (V) is a direct function of the exchange rate (R) and raceway length (L), very high R values can be required to achieve $V = 3$ cm/sec in rectangular raceways of practical length (Westers 1984). Water velocity in a linear raceway is given by:

$$V = LR/36$$

where L = length, R = water exchange rate and 36 = constant. For example, in a raceway with $R = 1$, the minimum length to obtain $V = 3$ cm/sec would be 108 m (354 ft).

In rectangular raceways, the flow required to obtain a given velocity can also be attained by proper dimensioning of the cross-sectional area (width and depth):

$$\text{Flow (gpm)} = VA(7.48)(60)$$

where V = velocity (ft/sec), A = cross sectional area of raceway (ft^2), 7.48 = gallons/ft^3, and 60 = sec/min. Thus, for a raceway 2 ft deep by 6 ft wide, the flow required to obtain a velocity of 3 cm/sec (0.09 ft/sec) would be:

$$\text{Flow} = (0.09 \text{ ft/sec})(12 \text{ ft}^2)(7.48 \text{ gal/ft}^3)(60 \text{ sec/min}) = 485 \text{ gpm}$$

In circular rearing units, water velocity is relatively independent of the exchange rate (R) because the water can be jetted in at various angles and under high pressure. Circular tanks are thus self-cleaning at lower values of R. Aquaculturists should keep in mind, however, that water velocities high enough to keep particulate matter suspended (self-cleaning tanks or raceways) may also cause chronic low level gill irritation predisposing the fish to bacterial gill disease. Chronic low level gill irritation is especially likely if the particles are in the 5–10 μm size range (Bullock et al. 1994).

Using impaired growth and food conversion as the index of a biological effect, maximum density index guidelines (lbs per ft^3 per inch of fish length) have been established for the raceway culture of a number of salmonids. These are 0.5 lbs/$ft^3 \cdot in^{-1}$ (3.4 kg/$m^3 \cdot$cm) for rainbow trout, 0.4 (2.7 kg/$m^3 \cdot cm^{-1}$) for coho salmon, and 0.3 (2.0 kg/$m^3 \cdot cm^{-1}$) for chinook salmon and cutthroat trout (Piper et al. 1982). Based on these guidelines, it appears that coho salmon require a minimum of about 1.3 times more living space than rainbow trout while chinook salmon require about 1.7 times as much. Density tolerance recommendations for the Atlantic salmon are complicated by the fact that parr of this species have a behavioral requirement for discrete feeding territories. That is, availability of bottom surface area is important as well as volumetric space. Failure to allow sufficient areal space (kg fish/m^2 bottom surface area) results in fin erosion, poor growth, and reduced survival. At cold (8–10°C) water temperatures, fin erosion can culminate in the partial or complete loss of the dorsal, pectoral, and pelvic fins. Fortunately, the damaged fin tissue can regenerate if the water temperature is increased.

At the 17–18°C temperatures commonly used by hatcheries that can heat water to accelerate growth, juvenile Atlantic salmon can apparently be reared to a final areal density of up to 41 kg/m^2 (229 kg/m^3 of rearing unit volume) without adverse effects on growth, survival, or fin condition (Soderberg et al. 1993). In colder water, (<10°C), areal densities as high as 21 kg/m^2 can be attained. Under production hatchery conditions however, much lower densities (such as 4–5 kg/m^2) may be more advisable. Many Atlantic salmon hatcheries use 5 kg/m^2 as a general guideline for parr. When the juveniles reach the smolt stage and begin to occupy more of the water column, density tolerance probably should be expressed in terms of kg/m^3. Here, less information is available although Mazur and Iwama (1993) have shown that densities as high as 32 kg/m^3 have no apparent effects on the health and physiological condition of juvenile Atlantic salmon smolts. Increasing the rearing density to 64 kg/m^3 results in hypercholesterolemia and significantly lower numbers of antibody-producing white blood cells.

Guidelines for the flow loadings (lbs fish/gpm flow) and densities (lbs/ft^3) commonly used to protect the health and physiological condition of Pacific anadromous salmonids reared in intensive culture are given in Table 4.2 as a function of fish size (Banks et al. 1979). These loadings have been designed to minimize metabolic stress by providing sufficient water flow to maintain the DO above 8 ppm, un-ionized ammonia concentrations below 0.01 ppm, and to minimize crowding stress by providing sufficient physical space to maintain the density factor at about 0.35 or below. Under production conditions, however, it is not uncommon for hatcheries to rear Pacific salmon at density factors of 0.45 or higher with no apparent effects on freshwater growth, health, or survival. As mentioned, raceway rearing densities that appear to be satisfactory during freshwater rearing may nevertheless result in effects on health and physiological condition that are eventually expressed as reduced ocean survival. At present, densities <10–15 kg/m^3 are suggested for spring (stream type) chinook salmon. Coho appear to be more resistant to higher raceway densities during rearing (Ewing and Ewing 1995). Nonanadromous salmonids such as rainbow trout, lake trout, and brown trout are more resistant to crowding stress and can tolerate density factors up to 0.5 without compromising health or physiological condition. However, flow rates should still be adjusted to maintain the DO >6 mg/L and NH_3 <0.01 mg/L to minimize metabolic stress. Guidelines for these species are given in Table 4.3. Similar guidelines for cool- and warmwater species such as northern pike, walleye, channel catfish, and striped bass are given in Tables 4.4 and 4.5. Again, the pathogen load of the water supply and the horizontal transmission of infectious agents are important considerations when adapting these guidelines for particular hatchery situations.

TABLE 4.2. Guidelines for maximum weight loadings (lbs/gpm inflow) and space densities (lbs/ft^3) to promote health, condition, and disease resistance of Pacific salmon in intensive raceway culture (data compiled from Banks et al. (1979).

Temperature		Fish Size (inches)						
°C	°F	1.5	2.0	2.5	3.0	4.0	5.0	6.0
				Coho or Fall Chinook Salmon (lbs/gpm)				
4.4	40	3.6	6.5	12.2	19.8	26.3	32.7	36.6
7.2	45	2.1	3.8	7.1	11.4	15.2	18.8	20.9
10	50	1.5	2.6	5.0	8.0	9.8	11.2	12.5
12.8	55	1.1	2.0	3.9	5.7	6.7	7.6	8.5
15.6	60	0.9	1.6	3.1	3.9	4.6	5.2	5.8
				Spring Chinook Salmon (lbs/gpm)				
4.4	40	2.7	4.9	9.1	14.6	19.7	24.5	27.5
7.2	45	1.6	2.9	5.3	8.6	11.4	14.1	15.7
10	50	1.1	2.0	3.8	6.0	7.4	8.4	9.4
12.8	55	0.8	1.5	2.8	4.3	5.0	5.7	6.4
				Steelhead Trout (lbs/gpm)				
4.4	40	3.6	6.5	12.2	14.8	26.3	31.8	35.4
7.2	45	2.1	3.8	7.1	8.6	15.2	18.2	20.2
10	50	1.5	2.6	5.0	6.0	9.5	10.8	12.0
12.8	55	1.1	2.0	3.9	4.6	6.5	7.4	8.2
				Chinook, Coho, Steelhead (lbs/ft^3)				
—	—	0.5	0.7	0.9	1.0	1.4	1.8	2.1

Notes: Loadings based on DO >7.5 mg/L, un-ionized ammonia <0.01 mg/L in raceway outflow, feeding rate <0.2 lbs/gpm/day, density factor <0.35. When adapting these guidelines to a particular hatchery situation, temperature, pathogen loads in the water supply, and the horizontal transmission of infectious agents must also be considered. Density factor is relatively independent of water temperature.

Since the high population densities used in intensive culture are also a primary factor in allowing fish diseases to develop and spread, managers can sometimes use the inverse as a tool. For example, salmonid fishes infected with bacterial gill disease (BGD) often recover spontaneously when the loading density is reduced (Bullock et al. 1994). Reducing population density not only improves environmental conditions for the fish, it also makes the horizontal transmission of pathogens (fish to fish) less probable. Reducing the population density can also help prevent indirectly caused diseases. For example, the highly intensified culture of channel catfish in cages may result in overbrowsing of natural sources of vitamins and essential nutrients within the cages. The fish develop characteristic fractures of the vertebral column just posterior to the rib cage. This so-called broken-back syndrome can be prevented by adding supplemental vitamin C to the diet.

TABLE 4.3. Guidelines for maximum weight loadings (lbs/gpm), and space densities (lbs/ft^3) in the intensive culture of lake, brown and rainbow trout. Minimum DO 6 ppm, 12°C

Fish Species	Fish Size (number per pound)						
	1000	500	100	50	33	25	15
	Weight Loadings (lbs/gpm)						
Lake trout	3.5	5.0	5.6	6.2	6.7	6.8	7.3
Brown trout	3.5	5.0	5.6	6.2	6.7	6.8	7.3
Rainbow trout	3.5	5.0	5.6	6.2	6.7	6.8	7.3
	Space Densities (lbs/ft^3)						
Lake trout	0.6	1.2	1.5	1.8	2.0	2.2	2.8
Brown trout	0.5	0.9	1.2	1.6	1.8	2.0	2.2
Rainbow trout	0.5	0.9	1.2	1.6	1.8	2.0	2.2

Source: compiled from IDC (1979).

Notes: Minimum DO = 6ppm. Pathogen loads in the water supply, and the horizontal transmission of infectious agents must be considered when adapting these guidelines to a particular hatchery situation.

Fish that are permitted to swim free in the same ponds can usually find ample vitamin sources to meet their needs. If the stocking density is too high to allow this, the diet provided can be designed to meet the specific nutritional needs dictated by the intensive rearing conditions used.

Because increasing the fish density to increase biomass production requires decreasing the fish loading to levels that often consume too much water, supplemental oxygenation is sometimes used as an alternative means to meet production requirements. Oxygen injection (usually from liquid oxygen) can easily maintain the DO at saturation or higher (hyperoxia) allowing both the fish loading to be increased and more water to be reused (Colt et al. 1991). Thus, fish production can be increased with no increase in water use or facility expansion. Blackburn and Clarke (1990) showed that densities as high as 25 kg/m^3 had no effect on the growth and smolting of zero-age coho salmon if DO levels were kept near saturation with oxygen injection. Caution should be used, however, in rearing salmonids under hyperoxic conditions for long periods of time because few guidelines exist for maximum safe DO concentrations. Rainbow trout reared at 130% O_2 saturation suffered decreased blood hemoglobin (Hb), hematocrit, and red blood cell count. Rearing at 150 % O_2 saturation decreased normal resistance to enteric redmouth disease (*Yersinia ruckeri*) (Caldwell and Hinshaw 1995). Growth is generally not increased under hyperoxic conditions and may decrease in some species. High blood CO_2 concentrations due to hypoventilation and the greater amount of

TABLE 4.4. Guidelines for maximum weight loadings (lbs/gpm) in the extensive culture of channel catfish and striped bass.

Temperature		Fish Size (inches)						
°C	°F	1.5	2.0	2.5	3.0	4.0	5.0	6.0
					Channel Catfish			
10	50	28	38	44	—	—	—	—
15	60	14	20	25	28	30	33	8
21	70	8	12	14	16	18	20	2
26	80	5	7	8	10	11	12	4
32	90	3	4	5	5	7	7	8
					Stripped Bass			
10	50	18	25	30	—	—	—	—
15	60	10	14	17	20	21	22	3
21	70	6	8	9	11	12	13	5
26	80	3	5	5	5	7	7	8
32	90	2	3	3	3	4	4	4

Source: Compiled from IDC (1979).

Note: Pathogen loads in the water supply and the horizontal transmission of infectious agents must be considered when adapting these guidelines to a particular hatchery situation.

TABLE 4.5. Guidelines for maximum weight density (lbs/ft^3) to minimize crowding stress and promote health, condition, and disease resistance in the extensive culture of cool- and warmwater fishes.

Fish Species	Fish Size (number per pound)						
	1000	500	100	50	33	25	15
Striped bass	0.38	0.65	0.80	1.0	1.2	1.3	1.6
Channel catfish	0.48	0.9	1.2	1.6	1.8	2.0	2.5
Walleye pike	0.55	0.9	1.2	1.3	1.5	1.8	2.0
Northern pike	0.40	0.8	1.0	1.4	1.6	1.8	2.0

Source: Compiled from IDC (1979).

Notes: Physiological tolerance to density (weight per unit volume) is relatively independent of water temperature. Pathogen loads in the water supply and the horizontal transmission of infectious agents must be considered when adapting these guidelines to a particular hatchery situation.

CO_2 from the respiration of the larger fish biomass elicit the release of catecholamine hormones that may vasoconstrict gut vasculature sufficiently to decrease nutrient absorption. Thus, although the amount of fish biomass produced per unit volume of water can be increased by O_2 injection, fish health, growth, and survival does not necessarily increase correspondingly and may decrease in some species.

TRANSPORTATION

Another source of stress associated with intensive (or extensive) aquaculture is fish transportation and distribution. Live fish have been transported, with varying degrees of success, since the origins of fish culture some 3,000 years ago in China and the Roman Empire. However, today's ability to economically transport large numbers of fish over long distances with almost routinely high survival rates dates from more recent times and is primarily based on the advances in fish physiology and improvements in the reliability of mechanized equipment that have occurred during the past 40 years. At present, fish transport and distribution operations are an indispensable part of aquaculture and fisheries management. Eyed salmonid eggs are routinely shipped around the world as are live tropical fish for the aquarium trade. Fry, fingerlings, and yearling fish are regularly trucked hundreds of miles for stocking into remote lakes and streams or even planted by air drop from aircraft. In warmwater aquaculture, farm-to-farm transfers of catfish, tilapia, and other species are routinely carried out. In salmon farming, juveniles are reared to smolt size in freshwater facilities on land and transported by helicopter, airplane, or ship to remote seawater net pen sites for grow-out. The massive Columbia River smolt-hauling operation conducted by the U.S. Army Corps of Engineers is a particularly impressive example of modern day fish transportation. Twenty million or more migrating juvenile coho, chinook, and sockeye salmon, and steelhead trout are collected each year in the upper Columbia and Snake rivers, barged or trucked hundreds of miles around the dams, and then released into the lower river to continue their seaward migration. Depending on the species, smolt survival immediately after release can approach 100% (Ceballos et al. 1991).

For the most part, the advances that have led to the successful fish transport equipment and methods used today have been driven by the economic necessity to move larger and larger numbers of fish in the smallest possible volume of water. Sophisticated life-support systems have been developed to meet the physiological needs of the transported fish and prevent the lethal water quality changes that would otherwise quickly occur. In the past, life-support systems have been developed mostly empirically. Today, life-support system design is more often based on the technical information on respiration, toxic excretory products, tolerance to stress, and water quality requirements considered in Chapters 2 and 3. A useful framework for a discussion of how such technical information is applied is to consider the basic criteria for life-support systems that will meet the physiological needs of transported fish, and specific transport methods and techniques that have been developed to mitigate stress and improve survival.

PHYSIOLOGICAL CRITERIA FOR LIFE-SUPPORT SYSTEMS

Although a variety of life-support equipment and methods has been developed over the years, the systems required to meet the physiological needs of transported fish are actually variations on only two basic types. These are (1) open systems in which fish are transported in continuously aerated tanks carried on trucks, trailers, aircraft, barges, or ships and (2) closed systems in which gametes, fertilized eggs, or (small) fish are transported in sealed containers, usually plastic bags. Regardless of type, dissolved oxygen must be provided and the accumulation of ammonia and carbon dioxide prevented.

Open system fish hauling tanks are usually elliptical or rectangular in cross section and insulated with an external layer (3–8 cm) of polyurethane or styrofoam for temperature control. Elliptical tanks, which are normally fabricated of aluminum or stainless steel, are the most expensive initially but they have a virtually unlimited service life. Their center of gravity is toward the center of the transport vehicle, and they weigh the least per unit volume of capacity. However, because of the lower initial cost, the great majority of transport tanks in use today are rectangular and constructed of fiberglass, aluminum, or marine grade plywood coated with fiberglass.

Although separate tanks that can be carried on existing trucks, trailers, ships, or aircraft are more versatile and the cost is substantially lower, custom designed vehicles dedicated to fish hauling are highly desirable. The fish hauling trucks and barges used by the U.S. Army Corps of Engineers dedicated to smolt transport operations around the dams on the Columbia and Snake Rivers are the largest and most technically advanced in current use. The truck- or trailer-mounted elliptical tanks hold 3,500 gallons (13,250 L) of water and incorporate life-support systems capable of supporting 1,700 pounds (771 kg) of smolts. Larger 150,000 gallon (568,000 L) barge-mounted rectangular tanks use pumped river water together with supplemental oxygenation to haul fish loads of about 75,000 pounds (34,000 kg). Species transported include migrating juvenile chinook, coho, and sockeye salmon, and steelhead trout. Hauling tanks used by state and federal fisheries agencies in routine hatchery operations in the United States are generally much smaller, averaging about 500 gallons (1,900 L) in size (Carmichael and Tomasso 1988).

Water Circulation. The most critical physiological requirement of hauling tank life-support systems, regardless of size, is the necessity to continuously circulate freshly aerated water to all parts of the container. This is absolutely essential to the health and survival of transported fish and a tank recirculation rate equivalent to at least 0.5 water exchanges per minute is widely considered desirable for both salmonids and warmwater fish. In large hauling tanks, water is usually recirculated through overhead spray bars by gasoline powered centrifugal pumps. However, centrifugal pumps cause compression and frictional heating of several degrees per hour and a mechanical refrigeration system must also be installed to maintain the hauling tank water temperature within the appropriate limits.

A simpler water circulation system consisting of 12-volt electrical agitators (aerators) mounted through holes in the tank surface is also coming into wide use. For shorter trips and smaller tanks (500–1,000 gallons), agitators have gradually become the dominant design because they are propeller driven and simultaneously circulate and spray-aerate the hauling tank water without the heating caused by centrifugal pumps. For larger tanks, agitators in combination with O_2 diffusers generally give better results. Exterior foam insulation (5–8 cm) painted white is generally adequate to prevent water temperatures from increasing more than 1°C per 4–5 hours of travel time in agitator equipped tanks, even in air temperatures of 30–35°C. For longer trips, chlorine-free ice can be added as necessary. One 12 volt agitator unit (of the Fresh Flow™ design) per 200 gallons of hauling tank volume will provide the recommended water recirculation rate of about 0.5 exchanges per minute. The 12 volt electrical power required can usually be provided simply by replacing the hauling truck's factory alternator with a commercially available 110-amp heavy duty unit.

The Aeration System. As discussed in Chapter 2, many species consume oxygen at 200–400 mg/kg/h even under normal conditions and this rate can double if the fish are excited, stressed, or are swimming rapidly. Because even a hauling tank as large as 1000 L contains at most only about 10 g of oxygen, it is obvious that a load of either cold- or warmwater fish large enough to be practical will consume most of it in only a few minutes. An aeration system to continuously replace the dissolved oxygen consumed by the fish is an absolute design requirement. Failure to provide sufficient dissolved oxygen quickly results in unconsciousness and death. However, an equally important function of the aeration system is to strip expired CO_2 from the water. This is often less appreciated.

Although the spraying action provided by recirculating pumps (or spray agitators) results in a small gain in DO, it serves primarily to air-equilibrate the dissolved waste gases and a supplemental aeration system is always required for larger tanks. This requirement is usually met either by injecting compressed air or oxygen into the intake side of the recirculating pump, if centrifugal pumps are used, or into diffusers mounted on the tank bottom.

The development of micropore carbon rod and ceramic diffusers has greatly increased the efficiency of gas transfer into the hauling tank water. Absorption efficiencies of 50% or more can now be attained and pure oxygen aeration systems have become economical. They are now the dominant design in fish transport systems. Because of the many variable factors involved, it is not practical to precalculate the oxygen flow rates needed to maintain the DO at or near saturation and also adequately strip out unwanted dissolved gases. The rate at which fish will consume the oxygen provided and produce CO_2 that must be removed depends on factors such as the degree of excitement and stress caused by the collection and loading procedures, swimming activity level, loading density in the hauling tank, time since last feeding, and water chemistry conditions such as temperature. The oxygen flow rate required will also depend on factors as disparate as the physical size of the bubbles produced by the diffusers, the concentration of dissolved CO_2 that must be removed, and the physical location of the diffuser in the tank. For example, if a perforated false bottom is used, the diffusers must be mounted above it or aeration efficiency will be greatly decreased. If carbon rod diffusers are employed, common practice by many state and federal fishery agencies is to install 0.3 meter of rod per Lpm of oxygen flow.

One potentially negative aspect of aeration with pure oxygen is the ease with which it becomes possible to supersaturate the hauling tank water. DO concentrations as high as 16–18 mg/L often quickly occur. Hauling fish in oxygen supersaturated water (hyperoxic conditions) does provide an economic advantage because either the total weight of fish hauled or the loading density safety margin can be increased. However, there is no real evidence of improved survival or other physiological benefit. A disadvantage of hyperoxia is that a marked elevation of CO_2 in the arterial blood occurs because the high DO reduces the normal gill ventilation rate,[10] increases the metabolic production of carbon dioxide, and decreases cardiac output. The blood buffering system compensates for the hypercapnia by increasing the plasma bicarbonate concentration corresponding (Hobe et al. 1984). When the fish are then released from the hauling truck, the elevated blood CO_2 rapidly diffuses into the receiving water where its concentration is much lower and the blood pH rises sharply. Physiological compensation for the resulting alkalosis may require as much as 24 hours. In the case of smolts transported in hyperoxic water and released directly into marine net pens, the sudden decline in blood gases includes oxygen as well as CO_2 because of the lower DO of seawater. The well-known phenomenon that smolts can become agitated when released directly into saltwater from freshwater hauling tanks may be due to the fact that their physiological systems are simultaneously being challenged with rapid changes in blood gases and blood pH as well as by the osmotic adjustments required by the rapid salinity change (Pennell 1991). Although these challenges are normally within the physiological tolerance limits of smolts in good condition, it is prudent to allow a recovery period of several days before subjecting the fish to the stress of other fish cultural procedures.

Transporting fish in oxygen supersaturated water also raises the possibility of gas bubble disease (GBD). As discussed in Chapter 3, the blood and tissues of fish swimming in supersaturated water quickly reach equilibrium with the higher partial pressure of the dissolved gases present and the bubbles that form in the vascular system can block circulation to vital organs. However, any O_2 bubbles

10. Fish respiration is regulated by the DO concentration rather than the dissolved CO_2 concentration as discussed in Chapter 2.

that form are quickly reduced in size by the normal metabolic demands of the tissues and mortality from oxygen GBD normally does not usually occur at supersaturation levels of less than about 200 %. However, the potential for GBD does exist when fish are transported by air. Because of the rapid reduction in atmospheric pressure as the aircraft climbs to its cruising altitude, the hauling tank water can easily become supersaturated with dissolved air. An altitude of only 1800 ft (0.6 km), will cause air supersaturation of 110% or more and bubbles may form in the fins, eyes, and yolk sac of transported fry and fingerlings (Hauck 1986). Larger fish are generally more resistant to supersaturation and GBD has not been reported to be a problem when smolts are transported by aircraft.

Preventive measures to be taken include minimizing the rate of ascent, flying at the lowest cruising altitude consistent with safety, chilling the hauling tank water to prevent any additional supersaturation caused by temperature increases, and rapid aeration to air-equilibrate the excess gases. If the stocking site is at sea level, the increase in air pressure during the descent may be sufficient to recompress gas bubbles that have formed and mortalities may not occur.

Experience has shown that aeration sufficient to provide a DO concentration acceptable for fish held under hatchery raceway conditions does not provide an adequate safety margin in transport operations because metabolism tends to be higher and more labile than normal due to excitement and stress. For salmonids, an aeration system that will maintain the DO concentration at 80% of saturation or above has become a routine requirement in hauling tank design (Westers 1984). This translates into a DO of about 10–12 mg/L if the hauling tank water is kept cooled to 10°C. The guideline for the oxygen flow rate needed to achieve this DO (for salmonids) is 3 Lpm O_2/250 kg of fish per meter of carbon rod diffuser length.

Ammonia toxicity. Salmonids and channel catfish excrete about 30 g of ammonia for each kg of food consumed and after these fish are loaded into a hauling tank, the ammonia (and other metabolic wastes) immediately begins to accumulate in the water instead of being flushed away. Depending on time since last feeding and the length of the trip, it is easily possible for the total ammonia concentration in a fish distribution tank to reach 10 mg/L or more (Smith 1978). As discussed in Chapter 3, the amount of the total ammonia in the toxic NH_3 form is the item of interest. This is primarily determined by the pH and to a lesser extent by the water temperature, total dissolved solids (TDS) concentration, and salinity. Chemical analysis measures the sum of both forms of ammonia present (NH_3 and NH_4^+), and the concentration of toxic NH_3 that the fish are being exposed to must be calculated. Selected values for the percentage of NH_3 that will be present over the range of pH and water temperatures of most interest in fish hauling operations in soft water areas were tabulated in Table 3.2. For transport operations in hard water areas or in seawater, the tabulated values will be 10–15% low and the standard correction for TDS or salinity should be applied.

In waters of low total hardness, the concentration of toxic NH_3 in the hauling tank will usually remain below the accepted 4-hour acute exposure limit of 0.1 mg/L even if the total ammonia concentration reaches 10 mg/L or more because the CO_2 that is also continuously being produced by the fish will usually maintain the pH at 7 or below.

Carbon dioxide toxicity. The accumulation of carbon dioxide in fish distribution tanks is generally a more serious problem than the accumulation of ammonia. Failure to adequately remove dissolved carbon dioxide results first in hypercapnia and acidosis, then respiratory stress from the Bohr and Root effects, tissue hypoxia, and eventually CO_2 narcosis and death. In general, fish produce about 1.4 mg of CO_2 for each mg of oxygen that they consume (Colt and Tchobanoglous 1981). Thus, a 1 kg load of fish consuming O_2 at 200 mg/h would also produce about 280 mg of CO_2 each hour. Rainbow trout transported at the normal loading density of 3 lb/gal (0.36 kg/L), can easily produce enough CO_2 to raise the initial concentration to an undesirable 30 mg/L or more within 30 minutes (Smith 1978). In practice, CO_2 concentrations of 20–30 mg/L occur within 30 minutes after loading most species unless steps are taken to prevent it by aeration and surface agitation.

As discussed in Chapter 2, the normal blood CO_2 concentration in salmonids and other active coldwater fish is quite low and the Bohr effect tends to be correspondingly large (begins at low blood CO_2 levels). Fish can tolerate moderately elevated dissolved CO_2 concentrations if the rate of increase is slow because the blood buffering system can compensate by increasing the plasma bicarbonate concentration sufficiently to stabilize the pH and prevent acidosis. Catecholamines may be released in response to the depressed cardiac output. However, CO_2 will accumulate faster than physiological compensation can occur in most fish transport situations. Concentrations greater than 15–20 mg CO_2/L are considered about the maximum for salmonids during hauling because respiratory stress from the Bohr and Root effects begins to occur at higher levels (Klontz 1992). If the dissolved CO_2 in the hauling tank water is allowed to increase above 20–30 mg/L, the blood CO_2 concentration will begin to rise and if the hypercapnia becomes severe, the blood pH will decline (respiratory acidosis) and oxygen transport to the tissues will again be impaired. In fishes with Root effect hemoglobins, acidosis will decrease oxygen transport to the tissues by decreasing the capacity of Hb to bind oxygen. For salmonids, the DO required to provide enough oxygen to the tissues to support a moderate swimming activity level increases from only 6 mg/L if little or no CO_2 is present to more than 11 mg/L if the dissolved CO_2 concentration rises to 30 mg/L (Basu 1959). Increased blood lactic acid concentrations (hyperlacticemia) because of excessive swimming activity can also cause (metabolic) acidosis and reduced oxygen transport to the tissues. One preventative treatment is to add millimolar quantities of sodium bicarbonate or sodium sulfate to the hauling tank water. This will offset the decrease in blood buffering capacity and help maintain a stable blood pH (Haswell et al. 1982). In contrast, warmwater fish can tolerate much higher CO_2 concentrations during transport because their blood hemoglobins usually exhibit a relatively small Bohr effect. For example, channel catfish can temporarily survive CO_2 levels or 50 mg/L or more during hauling if the DO is also near saturation (Tucker and Robinson 1990).

As previously mentioned, a major function of aeration is to prevent expired CO_2 from accumulating in the hauling tank water. The gas displacement and stripping action provided by the diffuser system helps transfer expired CO_2 from the hauling tank water into the tank headspace. Aeration with pure oxygen increases the efficiency of stripping CO_2 and other unwanted dissolved gases, such as N_2, from the hauling tank water. As oxygen bubbles from the aeration system rise through the water, the O_2 diffuses outward where its concentration is lower while dissolved CO_2 and N_2 diffuse inward and are carried to the surface and released into the tank headspace. Both processes are driven by the partial pressure differential of the respective gases. However, surface agitation is also required to complete the removal of CO_2 efficiently. In soft water areas, the slightly acid conditions promote the removal of CO_2 by reducing its solubility.

The efficient stripping and air-equilibration of dissolved CO_2 made possible by oxygen aeration and the surface agitation of spray aerators will accomplish little unless the CO_2 removed from the water is rapidly removed from the hauling tank headspace as well. Otherwise, it will simply redissolve. Merely removing the hauling tank covers is surprisingly effective. This has been shown to allow a dissolved CO_2 concentration of 30 mg/L to decrease to 5 mg/L in only a few minutes—with no change in aeration rate or other variable (Smith 1978). Obviously, a tank designed to have an open top would provide highly efficient CO_2 removal and such tanks are now being increasingly used to haul species that are not light sensitive. A 6–8 inch rim (15–20 cm) is installed around the top and bottom edges of the screened opening to minimize water loss from splashing. The open top design also allows some evaporative cooling to occur. When hauling smolts and other light sensitive fish, the traditional closed-top fish distribution unit must still be used. Covered tanks require forced ventilation of the headspace to continuously remove the CO_2 stripped from the water by the aeration system. Air scoops or vents mounted on the top surface can be used to provide the required airflow by ram ventilation while the hauling vehicle is in motion.

Respiratory stress from the Bohr and Root effects can be minimized by fish handling protocols designed to prevent hyperlacticemia (by minimizing excitement and swimming activity), and by an aeration system that will prevent hypercapnia by stripping out dissolved CO_2 as well as providing an adequate supply of dissolved oxygen. In practice, these are two of the most important considerations in meeting the physiological needs of transported fish.

A conservative guideline is that (salmonid) fish in distribution units will have adequate oxygen as long as the DO does not fall below about 80% of saturation and dissolved CO_2 levels are kept well below 30–40 mg/L.

TECHNIQUES TO MITIGATE STRESS AND IMPROVE SURVIVAL

Handling and Crowding. Haskell (1941) was apparently the first to correctly attribute the rapid drop in DO that occurs during the first half hour after fish are loaded into hauling tanks to the effect of excitement caused by handling on their rate of oxygen consumption. Since the DO concentration in hauling tank water can often be less than its normal maximum of about 10–12 mg/L, Haskell anticipated this initial oxygen demand by starting the aeration system 5 to 10 minutes before fish were loaded to ensure that the hauling tank water was initially oxygen saturated or supersaturated.

Today, it is known that the maximum weight of fish that can be safely loaded and transported in an open system fish distribution unit depends on the physiological requirements of the fish to be hauled, the efficiency of the aeration system, the water chemistry, and the hauling tank design. Species specific physiological characteristics, such as the corneal abrasion and blindness that can occur when largemouth bass are transported at densities that do not otherwise violate guidelines for recommended water quality, also strongly affect hauling tank carrying capacity (Ubels and Edelhauser 1987). Thus, the maximum loadings fish can tolerate must usually be determined by experience. In addition, loadings found to be successful for one tank design can be expected to be somewhat different from another even though their size is similar. Despite these limitations however, useful guidelines have been developed. For example, carrying capacity is known to be a direct function of fish size (Westers 1984). For salmonids, this maximum permissible weight is directly proportional to fish length. Thus, if experience shows that a particular distribution unit design will safely carry 100 kg of 2 cm trout, it will also be capable of transporting about 200 kg of 4 cm trout, all other factors being equal (Piper et al. 1982).

A complicating factor in recommending hauling tank carrying capacities (fish loading densities) for open transport system tanks is that loading density is commonly expressed in at least three different ways. Percent loading (weight of the fish as a percentage of the weight of the water), weight/volume loading (weight of fish per unit volume of water), or the displaced weight loading (weight of fish per unit volume of water minus the volume of water displaced by the fish) are all regularly used. In calculating loading densities by the latter method, the specific gravity of the fish is assumed to be 1.0 and volume of the water displaced (tabulated for convenience in Table 4.6 for nonmetric system users) is subtracted from

TABLE 4.6. Volume of hauling tank water displaced by a given weight of fish assuming an average value for the specific gravity of teleosts of 1.0 (calculated from information originally developed by McCraren and Jones (1978)).

Fish Weight		Water Displaced	
Pounds	Kilograms	Gallons	Liters
200	91	24	91
400	181	48	181
600	272	72	272
800	363	96	363
1,000	454	120	454
1,200	544	144	544
1,400	635	168	635
1,800	816	216	816

Note: It is now known that the specific gravity of Pacific salmonids can vary from 1.0 to 1.04 (Lewis et al. 1993). Thus, the values tabulated above may slightly underestimate the actual weight of water displaced.

the final volume since it is not available to the fish. The loading density (*LD*) is then:

$$LD = \text{fish weight}/(\text{water volume} - \text{water displaced by the fish})$$

Using this method of calculation, the loading density of 800 pounds of fish added to 500 gallons of water in a tank would be:

$$LD = 800/(500 - 96) \approx 2 \text{ lb/gal } (24\%, 0.24 \text{ kg/L}).$$

If the water displaced by the fish is not subtracted, the loading density would appear to be 800/500 = 1.6 lb/gal (19%, 0.19 kg/L), a substantially lower figure. As a first approximation, 2,000 lbs of fish will displace about 240 gallons of water—a significant amount. The water displacement method of expressing carrying capacity has been recommended by Piper et al. (1982), but any method can be used provided it is used consistently. However, it should be noted that the specific gravity of hatchery salmonids may actually vary from 1.0 to as high as 1.04, depending on the state of swim bladder inflation (Lewis et al. 1993). Thus, an inventory assuming a specific gravity of 1.0 when the actual value was 1.04 would underestimate the true weight by approximately 4%.

For nonanadromous rainbow trout, or brook, brown, and lake trout, the life-support systems now available allow hauling densities of about 36% (0.36 kg/L, 3 lb/gal) to be safely achieved when transporting larger (25 cm) fish (Piper et al. 1982; Westers 1984). For anadromous salmonid juveniles, carrying capacities in the range of 6–24% (0.06–0.24 kg/L, 0.51–2.0 lb/gal) for fingerlings, and 24–36% (0.24–0.36 kg/L, 2.0–3.0 lb/gal) for the larger parr (10–12 cm) are easily attained. In smolt-hauling, considerably lower loading densities are required because anadromous salmonids during this life stage are more sensitive to handling and crowding stress and scale loss (Barton and Iwama 1991). A weight loading of 5% (0.05 kg/L, 0.42 lb/gal) is a conservative guideline widely used for both Pacific and Atlantic salmon and steelhead trout. In barges or ships using pumped river or ocean water, the same weight loading with a flow rate of about 2 Lpm/kg of fish (0.2 gpm/lb) is used to achieve an adequate tank circulation rate (Ceballos et al. 1991).

Striped bass can tolerate hauling densities of 20% (0.2 kg/L; 1.6 lbs/gal) at a fish size of 100 grams (Mazik et al. 1991). Loadings up to 60% have been used for 100–200 gram channel catfish.

Anesthetics and Other Drugs. Anesthetic (loss of sensation) and hypnotic (sleep-inducing) drugs added to the hauling tank water can be helpful in improving the survival of transported fish. The main purpose of drug additives is to slow metabolic rates and thus reduce oxygen consumption and ammonia and carbon dioxide production. However, these agents can also mitigate physiological stress due to excitement and handling, and slow swimming activity as well. The latter is a significant help in preventing injuries due to excitement, such as broken fins, that would ordinarily occur during hauling, and in reducing scale loss—a serious cause of mortality when smolts are transferred directly into seawater. Using such drugs, rainbow trout have been successfully transported at two to three times the normal weight of fish per unit volume of water.

Of the many drugs tested as transport additives, the hypnotics sodium seconal, sodium amytal, quinaldine, etomidate, metomidate, and the anesthetics 2-phenoxyethanol (2-PE) and tricaine methanesulfonate (MS-222) have received the widest use (Wedemeyer 1996). Sodium seconal or sodium amytal used at 8 mg/L, or quinaldine at 2 mg/L reduce the swimming activity and oxygen consumption of a variety of species during hauling, but these compounds have not been cleared for use on fish for human consumption and so are illegal except for experimental use (Summerfelt and Smith 1990). Also, quinaldine at 2 mg/L does not effectively mitigate the corticosteroid stress response (Davis et al. 1982). Etomidate, the ethyl derivative of its parent compound propoxate, effectively controls adverse physiological changes in a variety of transported fishes when used at 0.1–0.2 mg/L—either with or without added 1% salt. At anesthetic concentrations (0.4–0.5 mg/L), etomidate can cause bradycardia, EKG abnormalities, and elevated blood pressure (Fredricks et al. 1993). However, at the reduced dosage used in transportation, etomidate treated fish suffer few physiological side effects and recovery is complete within a relatively short time after release (Amend et al. 1982; Davis et al. 1982).

Metomidate (Marinil®), the methyl derivative of propoxate, was originally developed as a short acting hypnotic for swine and is presently cleared (in Canada) for certain fisheries applications as well. The anesthetic dose for salmonids is 4–5 mg/L. As a transport additive for smolts, metomidate is used at the lightly sedating dosage of 0.1–0.2 mg/L (Pennell 1991). Treated fish will remain upright and respire normally, but they will generally not avoid netting and their corticosteroid stress response to handling will be greatly reduced (Kreiberg and Powell 1991). Metomidate has few adverse side effects and work is presently underway to license it for fisheries applications in the United States.

The anesthetics 2-PE and MS-222 have both been used as transport additives for a number of years. Although 2-PE provides the additional benefit of being a mild fungicide, prolonged skin contact may be harmful to human health and registration efforts to license it for routine use are unlikely. MS-222, originally synthesized as a local anesthetic for humans, is now approved for fisheries applications in several countries. Juvenile rainbow trout and other nonanadromous salmonids can be successfully transported sedated with MS-222 at about 25 mg/L (Piper et al. 1982). Higher concentrations (50–100 mg/L) are not recommended because they both anesthetize the fish and stimulate oxygen consumption. An undesirably long recovery period after release from the hauling truck is also required.

In smolt-hauling operations, MS-222 added at 25 mg/L causes loss of equilibrium and is not recommended. However, at 10 mg/L MS-222 provides significant physiological benefits as a transport additive. Equilibrium is not affected, and swimming activity and oxygen consumption both decrease; even at loading densities as high as 24% (0.24 kg/L, 2.0 lb/gal).

Although anesthetic or hypnotic drugs can make it possible to increase the fish loading density by a factor of two or three times the normal weight per unit volume of water, these agents should probably not be used solely for this purpose or to compensate for basic deficiencies in water quality (Piper et al. 1982). Drugs are legitimately used, however, to decrease the amount of scale loss resulting from excitement; a significant cause of mortality in smolts released directly into seawater (Bouck and Smith 1979). However, if drugs are to be used routinely, only compounds licensed for fisheries applications should be selected and proper dosages must be used. For anesthetics and hypnotics, the proper dosage is one that produces light sedation (reduced reaction to external stimuli without loss of equilibrium) rather than anesthesia (loss of sensation and usually, equilibrium). It is critical that equilibrium and essential physiological functions such as osmoregulation and gill respiratory exchange not be affected (Summerfelt and Smith 1990). Anesthetized fish that lose equilibrium will sink to the bottom of the hauling tank and may suffocate or be drawn against the pump screens causing scale loss and preventing adequate water circulation. As general precautions, a recovery period of several days should always be allowed before any drug is used a second time on transported fish and human skin contact should be minimized.

Fasting. Withholding food for a period of time prior to transportation is a well-known technique for improving the survival of transported fish. Fasting decreases both the amount of oxygen consumed and the amount of carbon dioxide, ammonia, and feces produced. However, because fishes are evolutionarily adapted to intermittent feeding, considerable time may be required for fasting to produce a noticeable physiological effect. It has long been known, for example, that the standard metabolic rate of rainbow trout does not begin to decline until about 48 hours without food. After about 60 hours of fasting, oxygen consumption decreases by about 25% and ammonia excretion by 50% (Phillips and Brockway 1954). In general, salmonids smaller than about 0.1 g must be starved for at least 2 days before they are transported and larger fish for at least 3 days in order to lower their metabolic rates enough to produce a practical effect.

At present, the recommended practice is to fast all salmonids for 48–72 hours before transportation.

Water Temperature. The importance of water temperature control has long been recognized as basic to improving the survival of transported fish. The most important aspects are (1) the use of reduced temperatures during hauling and (2) the temperature shock that can be caused by too large a mismatch between the hauling tank water and the receiving water.

The first records of the practical benefits of transporting fish in chilled water date to the 1800s (McCraren 1978). As is now understood, the benefits actually derive from the effect of water temperature on physiological processes. Thus, hypothermia reduces both oxygen consumption and the production of toxic excretory products. Because the Q_{10} for most physiological processes is about 2.0, reducing the water temperature by 10°C should reduce oxygen consumption and waste production during transportation by about 50%. However, the Q_{10} for the energy metabolism of salmonids actually averages somewhat lower (1.4–1.9) than the expected value of 2.0 and the relationship between energy metabolism and temperature is linear only over the range of 3–18°C (Smith 1977). Thus cooling the hauling tank water by only 8°C decreased the oxygen consumption of transported rainbow trout by about 50% whereas Meehan and Revet (1962) found that lowering the water temperature by 6°C reduced the oxygen consumption of transported sockeye salmon fry by only 20%.

In addition to reducing oxygen consumption, hypothermia also assists in mitigating aspects of the physiological stress response. Lowering the water temperature from 10°C to 5°C reduced hyperglycemia by about 30% in spring (stream type) chinook salmon smolts trucked at a loading density of 12% (0.12 kg/L, 1.0 lb/gal) (Wedemeyer 1996). Unfortunately, chilled water offers little protection against the blood electrolyte depletion that seems to be a major factor in the delayed mortality that can result from hauling smolts. Thus, the usefulness of hypothermia as a fish-transport control measure is probably limited to reducing oxygen consumption and ammonia production.

Within limits, cooling hauling tank water by 1°C will allow an increase in the fish loading density of about 10% (Piper et al. 1982). For this reason, hauling tank water is now usually cooled somewhat prior to loading fish as a matter of routine. Refrigeration systems or the use of chlorine-free ice to lower the water temperature to 5–7°C are currently the most widely used methods in salmonid transport operations. The amount of cooling needed can be judged from the difference between the hatchery water temperature and the temperature of the receiving water. However, the hauling tank water is generally not chilled more than about 10°C below ambient temperatures.

A mismatch between the hauling tank water and the receiving water can be stressful to transported fish. If the receiving water is more than 10°C colder than the hauling tank water, temperature shock can be minimized by acclimating the fish to the new water temperature over a period of several hours. The stress of temperature changes of less than 10°C is mild and usually well tolerated by salmonids. However, if the fish have been previously exposed to pathogens, the mild stress response that does occur may still be sufficient to activate latent infections such as furunculosis or bacterial kidney disease (Wedemeyer and Goodyear 1984).

Light Intensity. Reduced light intensity can be helpful in mitigating transport stress in anadromous salmonids. Spring (stream type) chinook salmon smolts subjected to a hauling challenge at twice the normal loading density of 5% (0.05 kg/L, 0.42 lb/gal) under darkened conditions showed a 25% reduction in the hyperglycemia and hypochloremia normally caused by handling and crowding (Wedemeyer 1996). For this reason, and because of the behavioral avoidance that occurs, taking steps to reduce light intensity during all stages of smolt transport is a recommended practice.

Mineral Salt Additions. When fish are stressed by handling and loading operations, the epinephrine produced increases blood circulation through the gills to increase the effective surface area and the amount of oxygen that can be obtained from the water. In freshwater, the normal osmotic influx of water through the gills rises substantially and urine production increases correspondingly to compensate. Life-threatening blood electrolyte imbalances can result from the loss of chloride and other ions in the resulting diuresis. The delayed mortality, commonly referred to as "hauling loss," which often occurs following transportation is at least partly due to this problem.

It has long been known that survival rates of transported cold- or warmwater fish could be substantially increased by simply adding NaCl at 0.5–1.0% to the hauling tank water. More recently, it has been shown that survival is even better if the fish can be allowed to recover in salt-enriched water after release from the hauling truck (Mazik et al. 1991). More complex mineral salt formulations have also been developed that are particularly useful in mitigating stress and reducing the mortality of fish transported in water of low total hardness (Wedemeyer 1996).

The physiological dysfunctions responsible for the immediate and delayed mortality that can occur when fish are transported, and the use of mineral salt additions to improve survival by mitigating these dysfunctions, have received considerable attention in recent years. The physiological benefits provided by either the simple or complex mineral salt formulations are probably mainly due to the protection they afford against the life-threatening blood electrolyte losses and ionoregulatory dysfunction that occur when the diuresis stimulated by handling and crowding stress is prolonged (Mazik et al. 1991). For example, the normal osmotic concentrations in soft water can be as low as 5 mOsmolar. Adding salt at 0.5% will increase this to about 300 mOsmolar; nearly isotonic with blood. However, salt formulations can also mitigate other adverse physiological changes, such as the depressed blood pH that results from metabolic acidosis in transported rainbow trout (Haswell et al. 1982), and the hyperglycemia and hypercortisolemia that occurs in transported anadromous salmonids (Wedemeyer 1996).

As judged from present experience, single or mixed formulations of NaCl, $CaCl_2$, Na_2SO_4, $NaHCO_3$, KCl, $MgSO_4$, K_3PO_4, and sea salts, used with or without tranquilizing concentrations of MS-222, have the most potential for alleviating the life-threatening physiological disturbances resulting from hauling smolts or other juvenile salmonids—especially in the slightly acidic waters of low total hardness typical of the west coast of North America. Two disadvantages of mineral salt additives are (1) the potential for uptake of Mg^{++} and K^{+} if scale loss of more than 10% has occurred, and (2) the potential for increased equipment corrosion problems (Wedemeyer 1996). A major advantage of drugs such as MS-222 as transport additives for smolts is the lack of potential for these two problems. A list of mineral salt formulations recommended to mitigate stress and improve the survival of transported fish is given in Table 4.7.

Polymer Formulations. Abrasions and scale loss due to handling and crowding are common problems in transported anadromous salmonids and other fish and the resulting infections that occur have long been known to be a significant cause of delayed mortality. Frequently, the cause of death is a secondary fungus infection—a common occurrence when even minor skin abrasions disrupt the normal protective slime layer and expose the underlying tissue to attack by *Saprolegnia.* Secondary fungus infections resulting from abrasions are also an important cause of prespawning mortality in returning adult Pacific salmon, such as the Sacramento River winter chinook, which must be trapped at dams, trucked to a hatchery, and held for extended periods before spawning. The tropical fish industry has successfully used water additives containing polyvinylpyrrolidone (PVP) or proprietary polymers to solve a related problem in transported aquarium fish. When abrasions and scale loss occur, these polymers temporarily bond to the exposed tissue to form a protective coating which is gradually sloughed off as healing takes place and the normal slime layer reforms. This technology has received

TABLE 4.7. Mineral salt formulations used for transporting Pacific salmon and steelhead smolts, rainbow trout, American shad, largemouth bass, and striped bass.

Salt Formulation	Concentration	Species Transported	Reference
$CaCl_2$	50 ppm	Chinook smolts	Wedemeyer 1996
$CaCl_2 + NaHCO_3$	50 + 250 ppm	Chinook smolts	Wedemeyer 1996
MS-222 + $NaHCO_3$	10 + 250 ppm	Chinook smolts	Wedemeyer 1996
MS-222 + $CaCl_2$	10 + 50 ppm	Chinook smolts	Wedemeyer 1996
MS-222 + $CaCl_2$ + $NaHCO_3$	10 + 50 + 250 ppm	Chinook smolts	Wedemeyer 1996
$NaHCO_3$, Na_2SO_4	10 Mm	Rainbow trout	Haswell et al. 1982
NaCl	0.1–1.0%	Rainbow trout, coho, chinook, steelhead	Carmichael and Tomasso 1988; Wedemeyer 1996
NaCl	1%	Striped bass	Mazik et al. 1991
KCl + NaCl + $MgSO_4$ + K_3PO_4	[a]	Largemouth bass	Carmichael et al. 1984
Sea salts	0.5%	American shad	Backman and Ross 1990

[a]Quantity sufficient to be isotonic with blood when added to minerals naturally present in the hauling tank water.

only limited use in cold- and warmwater fish culture where the approach has been to reduce mortality by using transport additives such as mineral salts and anesthetics that mitigate physiological stress. However, polymer formulations, usually added as one of the commercially available products such as Polyaqua®, are now being increasingly used by both the aquaculture industry and by state and federal conservation hatcheries as a water additive for transporting juvenile salmonids as well as other fish (Carmichael and Tomasso 1988). In the author's experience, a concentration of 100 ppm Polyaqua during trucking has significantly reduced the prespawning mortality of adult fall (ocean type) chinook salmon and steelhead trout caused by *Saprolegnia.* Results were also promising when Polyaqua was used at 100 ppm in the anesthetic and holding tanks while examining adult steelhead trout for spawning ripeness. At the U.S. Fish and Wildlife Service Coleman National Fish Hatchery, the prespawning mortality of treated adult steelhead subjected to repeated handling over a 3-month period was about 7% compared with the 30–50% mortality sometimes experienced.

TRANSPORT IN AIRCRAFT, SHIPS AND BARGES

Transporting fish with fixed-wing aircraft or helicopters can be very cost effective in some circumstances and it has now become a relatively common practice. Fixed-wing aircraft are used both to stock remote lakes and streams by air drop and to haul smolts to ocean net pen sites. Fish survival following air drop is known to be lower than with other forms of transport (Fraser and Beamish 1969), but smolt-hauling by float plane from freshwater hatcheries to remote seawater net pen sites gives good results.

In transporting smolts by float plane, custom designed tanks can be used, but plastic garbage cans lined with appropriately sized plastic bags save weight and also give good results. The bags are filled with oxygenated freshwater with or without added ice, depending on the length of the flight and the temperature differential between the freshwater hatchery and the sea pens. Generally, the smolts are loaded into the hauling containers at a density of 12–16% (0.12–0.16 kg/L). Loading densities at the lower end of the range are used for smaller fish. As with other forms of transport, salt (NaCl) at 0.1% is often added to mitigate stress. The cans are loaded onto the aircraft (usually a single or twin engine float plane), and the DO is adjusted and maintained at 10–12 mg/L with compressed oxygen during the flight. A typical flight is only 30–40 minutes but a backup supply of ice and oxygen should always be carried aboard the aircraft. If ice is not used, the water temperature may rise a few degrees. Upon arrival, the plastic bags containing the smolts are removed and floated in the sea pens allowing the temperature, salinity, and DO to gradually equalize.

Helicopter transport, although more expensive on a per hour basis, allows greater flexibility and has now become the dominant method of hauling smolts to marine farm sites by air (Shepherd and Bérézay 1987). The total weight of fish and water carried is determined by the lifting capacity of the helicopter used—generally about 400 kg (900 lbs). Transport containers can be carried internally but commercially available water carrying buckets designed for fire fighting that have been modified for fish hauling are more widely used. Available designs include rigid-wall fiberglass or aluminum, and collapsible fabric with a metal frame. A custom-designed aluminum fish hauling container with individual bottom dumping compartments opened by solenoids controlled from inside the helicopter has also been developed (Larson 1984). In either case, aeration is accomplished with a small compressed oxygen tank strapped to the container. A clear plastic lid on the bucket will minimize the chance of accidental dumps, reduce water loss, and allow the pilot to visually check the fish during the flight. If there is no ground crew at the release site, a bucket with a conical bottom will allow the pilot to overturn the container on a cage walkway to ensure that all fish are released (Shepherd and Bérézay 1987).

When large numbers of smolts or adults must be transferred long distances, tanks carried on ships or barges can be quite practical (Pennell 1991). Both smolt-hauling operations from shore-based hatcheries to remote net pen sites and farm-to-farm transfers of larger fish are carried out over distances that can be 1000 km (700 miles) or more. Smolts are usually transported in flow-through seawater tanks with capacities of 60,000 L (16,000 gal) or more at loading densities of 0.4–0.7% (4–7 kg/m^3). Supplemental aeration with oxygen or compressed air is used as needed and the condition of the fish is monitored with video cameras. The major precaution required is to avoid areas of toxic marine algae blooms.

Barges are being used extensively in the United States to transport migrating juvenile Pacific salmon and steelhead trout around the dams on the Columbia and Snake rivers in what is now the largest smolt-hauling operation in the world. Flow-through tanks as large as 570,000 L (150,000 gal) are used with liquid oxygen for aeration. River water is pumped into the tanks through spray bars or degassing columns for supplemental aeration and to eliminate supersaturation caused by the dams. The frictional heating generally caused by centrifugal pumps is of no concern in this application. Smolts are transported at densities of 5% (0.05 kg/L, 0.42 lb/gal). The water circulation rate is adjusted to about 2 Lpm inflow per kilogram of fish is used to achieve an exchange rate in the tanks of about 0.1 per minute ($R = 6$). The carrying capacity of a single barge is about 34,000 kg (75,000 lb). In some years, 20 million or more migrating juvenile coho, chinook, sockeye salmon, and steelhead trout are collected and transported around as many as eight dams and nearly 500 km of impounded river (Ceballos et al. 1991).

SHIPMENTS IN SEALED CONTAINERS

Gametes, fertilized eggs, alevins, and other small fish held in oxygenated closed containers cooled with crushed ice can easily be transported long distances. Salmonid eggs, tropical fish for the aquarium industry, juvenile salmonids, and warmwater fish used in aquaculture are routinely shipped around the world in this manner.

Gametes, Fertilized Eggs. Salmonid eggs can be transported during three stages of their development: as unfertilized eggs (gametes); as newly fertilized (green) eggs for about 12 hours after water hardening; and at the eyed stage (after the embryo has developed eye pigmentation).

Gametes are rarely transported. When it is necessary, the sperm is shipped separately in a sealed container (usually a plastic bag) on ice. An air space in the container of about 10 volumes of air to 1 volume of sperm is required to provide a sufficient amount of oxygen. Unfertilized eggs if shaken or jarred in transit will suffer high losses but can also be shipped in sealed containers on ice. To minimize movement, the containers are completely filled with eggs—no air space is allowed. If the temperature remains below about 5°C, the fertility of the transported sperm and eggs will remain high for at least 4–5 hours (Banks and Fowler 1982).

Newly fertilized (green) salmonid eggs can be shipped during the period of about 12 hours after water hardening but their initial sensitivity to mechanical shock begins to develop as soon as 15 minutes after fertilization, and precautions to prevent jarring or movement are absolutely essential (Banks and Fowler 1982). Shipping the eggs totally submersed in water with no air space in the container will help minimize losses. During the period from about 48 hours after fertilization until eye pigmentation has formed (eyed stage), incubating eggs should not be disturbed for any reason.

After eggs have developed to the eyed stage, they become resistant to mechanical shock and can be shipped out of water over long distances if kept moist and cool. Sealed shipping containers that provide the required environmental conditions have been designed and eyed eggs are now routinely transported around the world. The containers are commercially available and usually consist of an insulated box with stacked perforated trays arranged to keep the eggs moist and cool without actually immersing them in water. Wet cheese cloth may or may not be placed on the bottom of each perforated tray before the eyed eggs are poured in. If cheese cloth is used, it is then folded over the eggs and the trays are stacked vertically in the box. The top tray is filled with cubed or crushed ice to cool the eggs during shipping and to help retard hatching. As the ice melts, the cold water drips down through the perforated trays and over the eggs to keep them moist, facilitate gas exchange, and wash away toxic metabolites. To prevent cold shock, ice taken directly from a freezer at subzero temperature should never be used until it has warmed enough to begin to melt.

Alevins, Juvenile Fish. Small lots of alevins and other small fish can also be transported successfully in closed systems. Plastic bags partially filled with oxygenated water with an oxygen atmosphere are generally used. For fingerling salmonids, a loading density of 50–100 g/L with an oxygen:water ratio of at least 3:1 (v/v) is widely used for trips of 12 to 48 hours (Shepherd and Bérézay 1987). Tilapia (4–5 grams) are often shipped in plastic bags at densities up to 600 g/L. Suggested densities for shipping other cold- and warmwater fish in chilled sealed containers with an oxygen atmosphere are given in Table 4.8.

A number of successful protocols have been developed. One method, which allows up to five hundred 1–2 gram salmonids to easily withstand shipping times of 10 hours or more, employs milk dispenser bags commonly used in the dairy industry. The bags are made of double-walled food grade polyethylene, have a neoprene plug with a short length of self-sealing tubing attached, and are inexpensive. A bag holder is also commercially available. In this shipping method, the plastic bag is placed on the bag holder, the plug is removed, and about 1 kg (2.2 lb) of crushed chlorine-free ice is added through a funnel. About 10 L of oxygen-

TABLE 4.8. Loading densities (g/L) acceptable for shipping warm- and coldwater fish in cooled, sealed containers with an oxygen atmosphere.

	Shipping Time (hours)			
Type of Fish	1	12	24	48
Catfish				
fingerlings (8 cm)	100	75	50	25
fry (0.6 cm)	50	40	30	—
Aquarium species (2–5 cm)	100	75	50	25
Salmonids (1–2 g)	100	100	50	50

Source: Compiled from from information originally developed by Johnson (1979).

saturated chilled water is added followed by up to five hundred 1–2 gram fingerling salmonids previously acclimated to 5°C. The bag now containing fish, ice, and water is removed from the holder, the air squeezed out and the plug replaced. A hypodermic needle attached to an oxygen line is inserted into the self-sealing tubing and the bag partially inflated. A suitable shipping container is selected, such as an insulated plastic lined cardboard box or plastic bucket with tight fitting lid. The bag is placed inside and more ice is packed around it for additional cooling. If plastic buckets are used, the additional cooling can be achieved by adding water and freezing it before the plastic bag is inserted.

At the normal (5–10%) loading densities used, salmonid fish can be held in the cooled and oxygenated plastic bags for about 10–12 hours before CO_2 and ammonia begin to approach toxic levels. This time can be significantly extended by adding a small amount of zeolite or other commercially available ion exchanger to absorb the ammonia (e.g., clinoptiolite at 14 g/L), and tris buffer (0.02 M, pH 8) to prevent metabolic acidosis from the expired CO_2 (Amend et al. 1982; Stickney 1983).

FORMULATED DIETS

Together with the adverse effects on health of the fish cultural procedures just discussed, the formulated diets required in modern intensive fish culture can also cause physiological problems. By far the best understood of these are the nutritional imbalances and deficiency diseases of the salmonid fishes and the channel catfish. For example, trout, salmon, and catfish are inefficient users of dietary carbohydrate and liver glycogen deposits that may impair hepatic function will occur if nutritionally unbalanced rations are fed. Nephrosis and kidney malfunction can also occur. Diagnosis is by histologic examination of hematoxylin-eosin stained liver and kidney sections. In general, diseases due to nutritional imbalances will be infrequent if commercially formulated diets are fed.

Most other nutritional diseases in intensive fish culture are due to dietary deficiencies. A list of these, together with clinical signs useful in diagnosis, is given in Table 4.9. For the most part, deficiency diseases will also be rare if modern commercial diets that have been stored correctly are used. However, deficiency diseases may be encountered when alternative diet ingredients are being tested to solve

TABLE 4.9. Pathology associated with nutritional imbalances in the diets of cold- and warmwater fish important to aquaculture.

Fish Health Problem	Nutritional Association
Poor growth, skin darkening	Inadequate dietary protein, essential amino acids
Liver glycogen deposits, excessive visceral fat	Excessive carbohydrates
Anemia, swollen, pale liver, fatty infiltration	Inadequate dietary lipids, essential fatty acids
Lipoid liver degeneration, anemia	Oxidized lipids (rancidity)
Anemia, cloudy lens	Riboflavin deficiency
Impaired collagen formation, wound healing, hemorrhagic tendency, broken back syndrome in catfish	Ascorbic acid deficiency
Microcytic anemia	Folic acid deficiency
Clubbed gills	Pantothenic acid deficiency
Anemia, muscle degeneration, pancreatitis disease of Atlantic salmon	Vitamin E deficiency
Bilateral lenticular cataracts	Zinc deficiency
Increased susceptibility to infections: catfish, salmonids	Vitamin C deficiency, high dietary levels of copper
Nephrocalcinosis	Elevated dissolved CO_2, dietary selenium and calcium
Hepatocarcinoma (hepatoma); rainbow trout, tilapia	Aflatoxin contaminated feed ingredients
Glomerular nephritis, anorexia	Gossypol, cycloprene fatty acids in cottonseed meal
Sekoke disease (diabetes) of carp	Silkworm pupae as a diet ingredient

Source: Compiled from Roberts and Bullock (1989); Blazer (1992)

supply or economic problems, or when cultural methods are being developed for new species. For example, the culture of marine species such as juvenile Pacific halibut and intermediate temperature freshwater fishes such as northern and walleye pike is now of great interest but their dietary requirements are largely unknown. A complete treatment of fish nutrition including clinical signs of presently understood nutritional deficiency diseases can be found in Halver (1989).

ADVENTITIOUS TOXINS

Fish health problems can also be caused by a number of antimetabolites, toxins, and contaminants that are found either naturally or adventitiously in fish feeds. The physiological effects vary from mildly reduced growth to lethal toxicity. A few of these toxic materials occur as contaminants in the animal-derived commodities used as feed ingredients as a result of environmental pollution. Others, such as the nitrites and amines in fish by-products that can react during diet preparation or storage to form nitrosamines, occur naturally. The latter compounds are carcinogenic and low levels in formulated diets are not uncommon. Both dimethyl- and diethylnitrosamine have been shown to be carcinogenic in rainbow trout, however little else is known. Biologists should simply be aware that a potential for fish health problems exists.

Most of the adventitious toxins with the potential to cause fish health problems result from microbial growth on diet ingredients during storage and from the antimetabolites naturally found in plants. At present, only a few plant materials, such as soybean, cottonseed, and rapeseed (canola) meals, and corn and wheat products, are commonly used as diet ingredients. However, as animal proteins become increasingly expensive, new plant proteins will be increasingly sought as substitutes. Thus, the potential for fish health problems caused by previously unknown plant toxins can also be expected to increase.

Soybean meal is probably the most widely used of the plant materials presently incorporated into fish diets as a protein source. However, soybeans have long been known to contain a variety of antimetabolites and toxins that must be inactivated before use. These include trypsin inhibitor (TI); a protein that reduces growth by inhibiting the digestive enzyme trypsin; lectins, which can bind to free sugars or sugar residues on blood cell membranes causing hemagglutination; and phytic acid, which can cause nutritional deficiencies by binding required trace metals. Other antimetabolites with the potential to adversely affect fish health are probably also present in soybean meal.

Trypsin inhibitor is probably the most important of the antinutritional substances found in soybeans. It has long been known to be the major factor responsible for the depressed growth rate of fish and other animals fed improperly heated soybean products. In commercially prepared feeds, the TI in soybean meal is usually inactivated by heating or drying during processing. However, the required treatment conditions have been standardized for only a few fish species. Insufficient heating will leave residual TI activity that will interfere with protein digestion. Overheating the meal will also reduce growth by reducing the bioavailability of amino acids, particularly lysine.

Fish vary in their sensitivity to TI. For example, rainbow trout are highly sensitive and trypsin inhibitor should be virtually completely inactivated in soybean meals used for trout diets. Channel catfish can tolerate higher TI activities. However, at least 80% of the TI activity must be destroyed for (fingerling) channel catfish to grow at optimum rates (Wilson and Poe 1985).

Phytic acid (myoinosital hexaphosphate) is another of the naturally occurring toxic materials in soybean meals that can have adverse effects on fish health. Because phytic acid chelates required trace minerals such as calcium, magnesium, zinc, and phosphorous, micronutrient deficiencies are presumably responsible. For example, phytic acid in feeds at a concentration of 25 g/kg depresses growth and thyroid function and increases the incidence of cataracts in juvenile chinook salmon (Richardson et al. 1985). The reduced growth caused by phytic acid may be due to the fact that phytic acid binds with proteins to produce complexes that are resistant to digestion, as well as to the reduced bioavailability of trace metals (Spinelli et al. 1983). Cataract formation is presumably due to a zinc deficiency caused by phytic acid chelation. Dietary zinc is known to be essential for normal eye development. Zinc added at 0.05 g/kg to the above phytic acid containing diet prevented cataracts. Supplementing soybean-containing feeds with a mineral pack is now generally recognized as beneficial in preventing cataracts in salmonids. Originally it was thought that diets high in minerals contributed to the development of nephrocalcinosis, but now it appears that environmental factors such as high dissolved CO_2 levels together with diets deficient in minerals are more likely responsible (Ghittino 1989).

The lectins contained in soybeans are readily inactivated by heat treatment and also by fish digestive enzymes (pepsin) and therefore are not presently a source of fish health problems.

In summary, when the trypsin inhibitor and other antinutritional factors in soybeans are inactivated by proper heating, soybean meal has no adverse effects on the health of rainbow trout and catfish, particularly when supplemented with amino acids and minerals. However, soybean meal remains unsatisfactory for Pacific salmon, particularly fall (ocean type) and spring (stream type) chinook. Coho salmon can tolerate small amounts (Fowler 1980).

Another low-cost plant protein source for salmonid diets is rapeseed (canola) meal. With the increasing use of this plant, the potential for fish health problems due to its thioglucoside (glucosinolate) content has become more important. Antimetabolite glucosinolate compounds are relatively nontoxic themselves, but are readily hydrolyzed enzymatically and release goitrin, organic nitrates, and thiocyanates that can cause a number of adverse physiological changes. Thiocyanates cause decreased iodine uptake by the thyroid, but this is usually reversible by increased dietary iodine. Goitrin inhibits the synthesis of thyroid hormones resulting in reduced plasma thyroxin (T_4) levels and thyroid hyperplasia in rainbow trout. If the thyroid hyperplasia is sufficient to compensate for the antithyroid effect, T_4 levels and growth may be normal. Supplemental dietary iodine does not necessarily reverse the thyroid malfunction caused by goitrin.

To make rapeseed protein concentrate suitable as a protein source for fish, the antithyroid glucosinolates must be removed by water extraction, and heat treatment used to inactivate the hydrolytic enzymes. Low-glucosinolate varieties of the rapeseed plant (termed *canola*) have also been developed. Canola meals have been used in dry salmonid diets at up to 13% with no adverse effects on the growth rate, osmoregulatory capacity, or thyroid function of juvenile chinook salmon (Higgs et al. 1982). Low-glucosinolate canola meals given an additional treatment with heat and water can be used at up to 20% in Pacific salmon diets. However, this concentration of canola meal in rainbow trout diets adversely affects blood thyroxin levels and results in compensatory thyroid hyperplasia (Hardy and Sullivan 1983). To date, little is known of the effects of glucosinolates on very young fish, in which thyroid activity is higher, or on fish used as brood stock which would probably be exposed for several years.

Another potentially toxic constituent of the rapeseed plant is erucic acid, a C-22 monosaturated fatty acid. Juvenile coho salmon fed erucic acid at high dietary levels suffer atrophy of the kidney glomeruli and lipid accumulation in the epicardial connective tissue (Hendricks and Baily 1989). Fortunately, the amount of erucic acid in practical diets is usually negligible because commercially prepared rapeseed meal is solvent-extracted to reduce its oil content before it is sold as a protein supplement. Canola varieties have been bred to have a naturally low erucic acid content in their oil as well as a low glucosinolate content in the solid meal remaining after oil extraction.

Gossypol, a yellow phenolic compound that occurs in the pigment glands of the cotton plant, is a naturally occurring toxin that can cause serious fish (and animal) health problems when cottonseed meal is used as a plant protein supplement in feeds. Fortunately, heat and moisture can be used during processing to convert most of the gossypol to its bound form, which is nontoxic.

The effects of gossypol on fish health vary by age and by species.[11] Cottonseed meal is not commonly used in starter or brood fish diets. In grower diets for rainbow trout, the maximum recommended free gossypol concentration is 100 mg/kg (ppm). Growth suppression begins at concentrations of about 300 ppm. At free gossypol concentrations above 500 ppm, severe reductions in hematocrit, hemoglobin, and plasma proteins occur together with pathological changes in the liver

11. Humans are resistant to the toxic effects of gossypol.

and kidney such as glomerular nephritis, necrosis, and ceroid deposition. The tolerance of other salmonids, channel catfish, and tilapia to gossypol is variable, although high levels usually result in depressed growth and pathological changes as they do in rainbow trout. For example, chinook salmon grow normally on diets containing up to 34% glanded cottonseed meal, but coho salmon suffer growth depression at only 22% cottonseed meal (Fowler 1980). The toxicity of gossypol to other fish species is less well understood.

Cottonseed meal also contains toxic cyclopropene fatty acids (CPFA). CPFAs are synergists of the carcinogen aflatoxin B_1 (discussed later in this chapter) and may also be carcinogens themselves. The two most important CPFAs, sterculic and malvalic acid, occur in other oilseed plants as well. Fortunately, normal processing of cottonseed meal to remove the oil also removes most of the CPFA content. However, salmonid diets formulated with 20–30% cottonseed meal may also contain concentrations of 7–10 ppm of toxic sterculic acid. The pathological effects of dietary CPFAs on rainbow trout usually begin when sterculic acid concentrations exceed about 10 ppm. Pathological changes include altered lipid metabolism, hepatocyte necrosis, glycogen deposition, bile duct proliferation, fibrosis, and increased liver-to-body-weight ratios. Neoplasms can occur when methyl sterculate levels exceed 15 ppm (Hendricks and Baily 1989). Thus, the maximum 7–10 ppm concentration normally found in salmonid diets is probably marginally safe, particularly for older trout. However, broodstock can incorporate dietary CPFAs into body lipids that can then pass into the egg and into the developing embryo. Fry and fingerlings are generally susceptible and cottonseed meal containing CPFAs should probably not be used in starter diets. Comparatively little is known about the effects of CPFA in fish other than rainbow trout. The residual CPFA content of cottonseed meal can cause growth depression in channel catfish (Robinson et al. 1984) but sockeye salmon appear to be resistant.

A number of other natural toxins that potentially could affect the health of cultured fish also occur in plant materials. These include α-amylase inhibitors, alkaloids, favatoxins, cyanogenic glycosides, oxalates, toxic amino acids, and indoles. The α-amylase inhibitors present in wheat can interfere with carbohydrate digestion in both carnivorous and herbivorous fish if not inactivated. In practical diets, however, wheat as a diet ingredient has not been shown to cause problems. Similar considerations apply to the toxic alkaloid compounds that occur naturally in plants. Thousands of these nitrogen-rich compounds have been identified and their toxic effects (often neurological) characterized. Fortunately, alkaloids are produced only by angiosperm plants and the blue-green algae, neither of which are presently used in fish feeds. For example, the alkaloid anatoxin-a (discussed in Chapter 3) that is released into the water when blue-green algae cells die or are disrupted causes serious fish health problems but would never be found in a present-day feedstuff. The possibility does exist that terrestrial alkaloid-producing plants could contaminate mechanically harvested crops such as soybeans or cotton that are used as protein sources in fish diets. For example, Hendricks et al. (1981) showed that alkaloids from the tansy ragwort plant cause liver megalocytosis, occlusion of hepatic veins, and necrosis in rainbow trout. Although not pathognomonic, if this type of liver pathology is seen, the possibility of alkaloid exposure should probably at least be considered.

Little is known of the effects on fish of favotoxins, cyanogenic glycosides, oxalates, toxic amino acids, and other such compounds because the plants containing them are also not yet used in fish diets.

RANCIDITY

The polyunsaturated fish oils used in feeds readily react with atmospheric oxygen producing a variety of toxic aldehydes, ketones, peroxides, and acids through a process termed *oxidative rancidity*. The compounds produced are responsible for the off flavors and odors characteristic of rancidity and can cause serious fish health problems if they are ingested. The pathological effects of rancid diets include anemia, liver degeneration, growth depression and mortality. They are caused by both the direct toxicity of the oxidative products produced and the nutritional deficiencies that result when these oxidative products destroy required diet ingredients such as vitamins E and C. Antioxidants such as ethoxyquin, butylated hydroxytolyene (BHT), or the α-tochopherols (vitamin E) are routinely added to feeds to retard the development of oxidative rancidity. However, some degree of rancidity is unavoidable under practical conditions. Oxidative rancidity is assessed using the thiobarbituric acid test (TBA) or by measuring peroxide values (PV) using standard methods given by the Association of Official Analytical Chemists (AOAC). Freshly prepared diets have low or near zero TBA and PV values. Diets with TBA values of 50–75 (mg malonaldehyde/kg feed) and peroxide values of 5–10 (meq peroxide/kg feed) are considered slightly oxidized. Unfortunately, it is not yet possible to state maximum levels of oxidative rancidity that will not affect the health of salmon or other species. However, fingerling salmonids (and perhaps other species as well) seem to be more resistant than fry. For example, first-feeding coho and Atlantic salmon (and probably other species as well) are sensitive to even low levels of oxidative rancidity as measure by TBA and PV values of 5–20 (Ketola et al. 1989). To minimize problems with oxidative rancidity, fish feeds should always be stored at the recommended temperature and used within the recommended time period, usually about 90 days for dry diets. Moist pellets should be stored frozen and thawed only in time to be fed. Frozen storage should not exceed about 6 months.

MOLD TOXINS

An extremely important group of adventitious toxins are produced by the molds that can grow on improperly stored plant materials that are to be used as diet ingredients. The best known of these are the carcinogenic aflatoxins produced by the mold *Aspergillus* spp., which grows on cottonseed and peanut meal, corn products, wheat and rice, and other plant materials stored under damp conditions. Aflatoxins at concentrations up to several hundred μg/kg can be produced. To date, aflatoxins produced by *Aspergillus flavus* have been of most importance in fish culture, although toxins produced by other molds can also be a source of fish health problems.

Of the commercially important fish used in aquaculture, rainbow trout are probably the most sensitive to the toxic and carcinogenic effects of aflatoxins. Diets containing as little as 0.4 ppb will result in reduced growth, anemia, and liver neoplasms within a year. Brook trout, cutthroat trout, and brown trout are somewhat sensitive to aflatoxins, but are more resistant than rainbow trout. Warmwater fish are comparatively resistant although liver neoplasms have been report in the channel catfish. Coho, chinook, and sockeye salmon are very resistant to both the toxic and carcinogenic effects of the aflatoxins.

Mold growth can contaminate improperly stored feed ingredients with other mycotoxins as well. For example, *Aspergillus* and *Fusarium* spp. growing on the corn and wheat that are becoming relatively common ingredients in fish diets produce a variety of toxic compounds the best known of which are ochratoxin A and vomitoxin. The safe exposure level to vomitoxin is probably in the low ppb range for rainbow trout (Woodward et al. 1983). Safe levels for the dietary intake of ochratoxin A are unknown but the intraperitoneal LD_{50} is about 5 mg/kg in rainbow trout (Hendricks and Baily 1989).

As animal proteins used in diets become more costly, economic pressures to develop new plant feedstuffs as substitutes will increase as well and more mold-produced toxins will undoubtedly be discovered. For the foreseeable future, knowledge of mycotoxins will continue to be important in managing the health of fish in intensive culture.

CONTAMINANTS

Toxic organic and inorganic chemicals resulting from the bioaccumulation of environmental pollutants up the food chain can occasionally find their way into fish feeds. Wild fish, such as the herring used to make fish meal, can be contaminated with such toxins both through the ingestion of contaminated food organisms and uptake from polluted waters and sediments. By way of contrast, dietary intake is usually the major route of exposure for fish in intensive culture because steps are normally taken to protect water quality and sediments are normally not present. At present, information on the health effects of organic and inorganic contaminants in diets is limited mostly to organochlorine pesticides, polychlorinated biphenyls (PCB), polycyclic aromatic hydrocarbons (PAH), heavy metals, and selenium.

The organochlorine pesticides of most potential concern are DDT, toxaphene, and dieldrin. However, these compounds have not been used in the United States for many years and their environmental concentrations have slowly decreased. It is now unlikely that feed ingredients produced here could be sufficiently contaminated to be a source of fish health problems. Unfortunately, chlorinated hydrocarbon pesticides are still widely used in other parts of the world where they persist in the aquatic environment and can accumulate in food chain organisms consumed by fish. Generally speaking, dietary concentrations in the low ppm range are required for toxicity. Clinical signs can include excitability and erratic swimming behavior due to disruption of nervous system function, a variety of pathological changes in the liver and kidneys, and inhibition of gill, kidney, and brain ATPase activity. Unfortunately, the acute and chronic pathological effects are usually nonspecific. The broken-back syndrome in trout and other fish exposed to water-borne toxaphene is pathognomonic but does not occur as a result of dietary exposure.

PCBs are also no longer used in the United States, but these compounds are extremely resistant to biodegradation and have accumulated through the food chain to the point that virtually all marine fish oils, and thus commercial diets incorporating these oils, still contain small (<1 ppm) amounts (Leatherland and Sonstegard 1984). Prolonged feeding of PCB-contaminated diets to brood fish can result PCB concentrations in eggs (7 ppm) sufficient to cause salmonid fry mortality. Fortunately, dietary PCB concentrations must be fairly high (50–100 ppm) to affect growth rates, although lower concentrations will induce the hepatic mixed-function oxidase enzyme system (Voss et al. 1982). In the United States, contamination of commercially formulated hatchery diets with adventitious PCBs should be rare if high-quality fish oils and other dietary ingredients from aquatic sources are used.

Dietary exposure to PAH compounds is known to be harmful to fish health but high concentrations and prolonged feeding are required. For example, rainbow trout fed 1000 ppm benzo[*a*]pyrene for 18 months suffered reduced growth, induction of liver mixed-function oxidase enzyme activity, and a 25% incidence of hepatocellular carcinomas (Hendricks et al. 1985). However, PAH levels in diets commercially formulated in the United States are normally orders of magnitude lower and again, fish health problems should usually not be encountered.

Heavy metal contaminated diets are also an infrequent occurrence, although organic mercury compounds have occasionally been found in fish by-products used as diet ingredients. Fortunately, the fish toxicity of dietary organomercury compounds is fairly low if tissue concentrations accumulate slowly. For example, methylmercury fed at concentrations of 24 ppm for 3 months caused gill hyperplasia and elevated hematocrits in rainbow trout but no mortalities (Wobeser 1975). However, muscle tissue concentrations reached 30 ppm, far in excess of the 0.5 ppm considered safe for human consumption. More recent work has shown that one sublethal effect of dietary methylmercury (fed at 2.5 μg/g fish) on fish health is to damage the sensory epithelium of the inner ear (Skak and Baatrup 1993). This may explain the abnormal swimming behavior and loss of equilibrium sometimes seen in mercury-exposed fish. Little is known about effects of dietary exposure to other heavy metals. Rainbow trout feeding on benthic invertebrates contaminated with cadmium, copper, and lead accumulated these metals and suffered liver pathology, reduced growth, and some mortality (Woodward et al. 1993). Rainbow trout apparently do not absorb significant amounts of dietary lead (Hodson et al. 1978).

EFFECTS ON SMOLT DEVELOPMENT

A continuing problem in the efficient use of intensively produced fish for the conservation of wild salmon populations or for the production of food by commercial aquaculture has been the health, quality, and seawater survival of the juvenile fish produced. Although hatchery fish have substantially better survival rates through hatching and early development, the ocean survival of hatchery smolts released into rivers is often below the survival of naturally produced smolts.

One of the most important causes of reduced seawater survival is incomplete smolt development. In many cases, apparently healthy hatchery smolts though large and silvery, are inadequately developed physiologically and are unable to continue to grow and develop normally in full strength seawater. This condition may be difficult to detect during freshwater rearing because the incomplete parr-smolt transformation is often not apparent until after release. It appears first as impaired migratory behavior and seawater tolerance, and later as reduced early marine growth and survival. The failure of hatcheries to produce smolts in optimum physiological condition probably arises both from an incomplete understanding of the normal parr–smolt transformation, and from incomplete data on how rearing conditions such as raceway densities, unnaturally constant water temperature, artificial diets, and night lighting affect physiological development. Knowledge of all the physical, chemical, and biological factors that affect smolt survival is needed to provide the rationale for modifying hatchery rearing conditions and release strategies.

Experience has shown that within limits, smolt survival of most anadromous fishes tends to be directly proportional to size at release. Accordingly, the tendency has been to release large smolts in an attempt to increase contribution to the fishery and returns to the hatchery. Unfortunately, there are many case histories of released juvenile which, although clinically healthy and large enough to be smolts, have not completely developed physiologically. In extreme cases, these fish simply remain in the stream during the summer following release.

Hatchery procedures with particular potential for adverse effects on smolt development include photoperiod (day length) alterations, elevated water temperatures to accelerate growth, elevated loading densities, and prerelease disease treatments. Photoperiod is a very important environmental cue that can be distorted easily. For example, Pacific or Atlantic salmon smolts reared indoors under constant artificial light do not show normal migratory behavior. In contrast, smolts in shallow raceways under direct sunlight can develop silvery coloration and show migratory behavior as early as February. However, exposure to direct sunlight may also cause dorsal epidermal lesions and mortality, a condition commonly termed *sunburn*. Thus, both the photoperiod and the intensity of light must be managed to prevent adverse effects on smolt quality.

The seasonal cycle of smolt development in juvenile Pacific and Atlantic salmon has a strong endogenous rhythm, which is coordinated by the yearly seasonal changes in day length and water temperature. The required link between changes in the external environment and physiological changes in the fish is provided by the endocrine system. Water temperature controls the rate of physiological response to photoperiod so that aspects of smolt development such as migratory behavior are apparent sooner at warmer temperatures. For example, holding fall (ocean type) chinook salmon over the winter in hatcheries using ground water accelerates growth but the unnaturally high temperatures can distort the normal development of springtime seawater tolerance. Prolonged exposure to an unnaturally long photoperiod also inhibits growth and smolting. Similarly, either continuous short photoperiods or complete darkness will cause asynchronous smolting. The rate of change of photoperiod is also an important cue, and modifying the timing of smolting is usually more successful if a shift in phase of the photoperiod cycle is used, rather than a change in its frequency. In both Atlantic and Pacific salmon, the freshwater rearing period can be shortened by several months if an increasing or long photoperiod is applied during winter. Photoperiod also has a marked effect on the physiological quality of the smolts produced. Again, the direction and rate of change of day length are more important photoperiod cues than day length per se. After initial seawater entry, however, growth and survival are not further influenced by photoperiod. As one example, Clarke et al. (1981) originally showed that presmolts held for 2 weeks under a 17-h light, 7-h dark photoperiod (17L:7D) had significantly higher gill ATPase and plasma thyroxine (T_4) levels and were better able to regulate plasma Na^+ than fish held under an 8L:16D photoperiod. After about 7 days in seawater, however, there were no further differences between the groups.

Although photoperiod regulation to produce out-of-season smolts is not widely used in commercial salmon aquaculture, it has been used to control spawning in rainbow trout to ensure supplies of out-of-season eggs (Thrush et al. 1994).

Water quality is another important environmental factor in which seemingly minor alterations that cause no apparent adverse effects on health during freshwater rearing can result in normal appearing smolts with inhibited migratory behavior, decreased seawater tolerance, and increased susceptibility to disease. Sublethal exposure to heavy metals is particularly insidious in this regard. For example, copper exposure at only 30 ppb during freshwater rearing inhibits the normal development of gill ATPase but has no effect on growth or survival. However, normal migratory behavior after release is inhibited and seawater tolerance and preference is compromised. Herbicides such as 2,4-D that are sometimes used in intensive forest and range management and in agriculture can also reduce the seawater tolerance and survival of anadromous salmonids exposed during freshwater rearing. Dimethylamine salts of 2,4-D used for brush, weed, and vine control on noncrop lands, and for control of aquatic vegetation, can be particular problems in this regard (Wedemeyer 1982).

Elevated water temperatures to accelerate growth can also have adverse effects on smolt development. In hatcheries with water reuse systems it is sometimes practical to heat water to higher temperatures to increase growth rates. However care must be taken not to inadvertently delay or accelerate normal smolt development or the onset of parr-reversion. For example, a water temperature above 15°C significantly increases the growth rate of juvenile coho salmon, and zero-age smolts can therefore be produced in less than 1 year (Brannon et al. 1982). Unfortunately, warmer water may also impair the normal development of gill ATPase activity and other aspects of smoltification and accelerate the rate of parr-reversion. All these physiological effects can reduce seawater survival and have caused considerable economic losses in net pen salmon aquaculture operations (Saxton et al. 1983). If a delayed photoperiod cycle is employed to produce underyearling coho salmon smolts, water temperatures as low as 8°C can be used and this problem prevented (Clarke and Shelbourne 1986). In the case of fall (ocean type) chinook salmon, the accelerated parr-reversion that occurs when elevated water temperatures are used to increase growth can be prevented or minimized by rearing the juveniles at a salinity of 10–20‰ (Clarke et al. 1981). For steelhead, if elevated temperatures above 13°C are used to produce underyearling smolts, a period in cooler water 1 to 2 months before seawater entry will allow recovery of normal gill ATPase activity.

Several of the problems with smolt performance that have been noted to result from the economic requirements for intensive, rather than extensive, fish culture for anadromous salmonids have led to the belief that improvements in hatchery practices can be made that will contribute substantially to increased ocean survival and returns to the fishery. Two areas of particular interest in this regard are the association of raceway rearing densities with subsequent ocean survival and the side effects of prerelease fish disease treatments given to control parasite and bacterial infections.

Good fish health management requires that smolts released from freshwater hatcheries be free of external parasite infestations and bacterial infections. Gill *Ichtyobodo* infestations, for example, result in impaired ability to osmoregulate in seawater because of physical damage to the chloride cells. However, some of the drugs and chemicals used for prerelease disease treatments can also impair osmoregulatory ability, resulting in the release of disease-free smolts unable to survive in seawater. Particularly poor seawater survival occurs if coho or chinook salmon smolts are treated with copper sulfate, Hyamine 1622, potassium permanganate, malachite green, or MS-222 prior to transfer into 28 ppt seawater (Bouck and Johnson 1979; Smith et al. 1987). Smolts treated with formalin and nifurpirinol suffer about a 10% mortality while trichlorfon or simazine disease treatments cause low or no mortality. If a 4-day freshwater recovery period is allowed, the seawater survival of fish treated with copper sulfate, potassium permanganate, MS-222, and malachite green is improved; the seawater mortality rate of hyamine 1622 treated fish is much reduced and becomes nil for the other agents. A 2-day recovery period is apparently adequate for juvenile fall chinook salmon given standard permanganate treatments (Smith et al. 1995). These findings, which are summarized in Table 4.10, suggest that it would be prudent to allow at least a 1-week freshwater recovery period before smolts given any fish disease treatment are transferred into seawater. In the case of conservation hatcheries, the river travel time of released smolts may provide the required recovery period. In addition, it should be kept in mind that the disease treatment may improve overall fish survival more than enough to compensate for the loses caused by the treatment chemicals used.

A second intensive rearing practice that can reduce the ocean survival of anadromous (Pacific) salmon is raceway densities that are too high. Fish densities are sometimes increased from the commonly used 10–15 kg/m^3 into the 30–40 kg/m^3 range as a means of increasing smolt production (Banks 1994). Although proportionally increased numbers of coho and chinook smolts can usually be released if raceway rearing densities are increased, it has been known for some time that the ocean survival of adults, particularly of spring chinook salmon, may be considerably reduced. The ocean survival of adult coho salmon seems to be much less affected by raceway rearing density over the range of about 15–40 kg/m^3, although Banks (1994) has shown that physical crowding can still have some effect on subsequent adult returns. The specific relationships between rearing density and ocean survival are as yet not well understood (Ewing and Ewing 1995). Extensive research has revealed few effects on condition factor, thyroid hormones, or other commonly measured aspects of smolt development that could explain the reduced adult survival. However, reduced immune competence and resistance to

TABLE 4.10. Effect of treatments with commonly used drugs and chemicals on the 10-day survival of coho salmon smolts in 28 ‰ seawater.

Fish Disease Therapeutant	Treatment Regimen			Seawater Mortality (%)	
	Dose (mg/L)	Time (min)	Days Treated	Direct Transfer	4-day Recovery
Copper sulfate	37	20	1	100	20
Endothal	5	60	1	100	4
MS-222	100	6	1	100	12
Potassium permanganate	2	60	1	80	12
Hyamine 1622	2	60	1	68	4
Malachite green	1	60	1	44	12
Oxytetracycline	1	60	1	20	12
Formalin	200	60	1	10	0
Nifurpyrinol	1.5	60	4	8	0
Simazine	2.5	60	1	4	0
Trichorofon	0.5	60	1	0	0

Source: Compiled from information originally developed by Bouck and Johnson (1979); Smith et al. (1987).

Notes: A 4-day recovery period after treatment greatly increases seawater tolerance. A 2-day recovery period is adequate for juvenile fall chinook given standard potassium permanganate treatments (Smith et al. 1995).

Vibrio anguillarum have been found in coho salmon reared at high (0.45 $lbs/ft^3 \cdot in$) density factors[12]; equivalent to about 45 kg/m^3 at time of release (Schreck et al. 1985; Maule et al. 1989). Gill ATPase activity is enhanced if coho salmon are reared at 14 kg/m^3 instead of the more usual 27 kg/m^3 (Sower and Fawcett 1991). In addition, favorable or unfavorable ocean conditions (such as poor upwelling) may partially mask the effects of high rearing densities on adult survival. Finally, the greater numbers of smolts released can still result in larger absolute numbers of adult returns if the percentage decrease in ocean survival due to high density rearing is not excessive. However, presumptive evidence exists that for spring chinook, at least, hatcheries operated at low or intermediate raceway loading densities (10–30 kg/m^3) may yield as great an adult return as hatcheries operating at their maximum capacity (Banks 1994). In other words, the higher feed and other production costs associated with high rearing densities may not be offset by correspondingly higher adult salmon production.

Guidelines are needed for hatchery management strategies that will help ensure maximum seawater growth and survival. On the basis of our present under-

12. As noted in Table 4.2, a maximum density factor of 0.35 is recommended. This figure is often exceeded by production hatcheries.

standing of the effects of rearing conditions on the coordinated physiological processes that occur during smolt development, the following suggestions, modified from those originally given by Wedemeyer et al. (1980), are presented.

- Water temperature should follow a natural seasonal pattern. If elevated temperatures are used to promote growth, this is best done during October through December. Temperature should not be elevated too quickly or to more than about 10°C in late winter unless accelerated smolting is desired. If possible, water temperatures should be held below 13°C at least 60 days prior to release of Atlantic salmon and steelhead trout, and below about 12°C for coho and chinook salmon to prevent premature smolting and desmoltification. The specific temperature regimens needed to produce the most successful smolts should be determined for each hatchery.
- If genetic strains are similar, and in the absence of complicating factors such as altered river and estuarine ecology, smolt releases should be timed to coincide as nearly as possible with the historical seaward migration of naturally produced fish. At headwater production sites, earlier release of hatchery fish may be called for. Planting large parr in the fall may be reasonable if observation shows naturally produced parr move to downriver sites at that time. The desired result is that hatchery-reared smolts that are genetically similar to wild smolts enter the sea at or near the same time.
- Proper photoperiod regulation is one of the most important environmental factors affecting production of functional smolts. Although the facilities needed for photoperiod control are minimal, this environmental priming factor should be altered with caution, since it is easily misused with disastrous results. Unless accelerated or delayed smolting is required, the best procedure is to hold the presmolts in outdoor ponds with no artificial light. This may preclude the use of bright floodlighting often used for night security. Indoor rearing should also be done under natural light if possible. Otherwise, artificial lighting should be timed to simulate natural intensity and photoperiod.
- Smolts must not be allowed to enter seawater with latent infections or gill parasite infestations. Ideally, prerelease disease treatments should be limited to medication known to have no effect on smolt performance. When this is not possible, a 7–14 day freshwater recovery period should be allowed before release.
- Physiological testing for osmoregulatory competence should be used to track smolt development and monitor effects of hatchery practices and environmental factors. Gill ATPase, plasma thyroxine (T_4), or the seawater challenge/blood sodium test are recommended.

Time, Age, and Size at Release

Hatchery practices that will optimize ocean survival and total returns to the fishery must center around the production and release of functional, healthy smolts at the most favorable size, age, and time. Historically, hatchery releases have been timed to mimic the migration of resident wild salmonids in the rivers in question. However, in many waterways there have been mild to severe changes in the aquatic environment that now render historical migration times inappropriate. These changes include the occurrence of gas supersaturation in the spring, restrictions on stream flows due to the requirements of intensive agricultural and hydropower generation, and fluctuations in temperature and food supplies in river, estuarine, and coastal areas. In addition to these factors, the genetic composition of the hatchery fish in any particular river may have been changed significantly from that of the wild type due to the practice of importing eggs from stocks in other watersheds and the (sometimes inadvertent) genetic selection that occurs during hatchery egg-taking operations. Superimposed on these problems are the constantly evolving needs of resource management programs, such as the recent development of ocean ranching, or delayed releases used to establish resident salmon populations in local marine areas.

Numerous studies attempting to quantify the effects of smolt size at time of seawater entry on subsequent ocean survival have been conducted. For example, hatchery-reared Atlantic salmon typically develop a bimodal size distribution. Fish from the larger size group (termed *upper modal*) can be transferred directly into full strength seawater net pens where they will grow from their initial size of 50–60 grams to 3–4 kg in about 18 months. Fish from the lower size mode (typically <30 grams) have high seawater mortality and the survivors rarely grow normally. Peterson (1973) was one of the first to show that releasing 2-year-old Atlantic salmon smolts smaller than 14 cm (i.e., lower modal fish) gave low adult returns. In contrast, releasing yearling smolts at lengths of 12–14 cm (upper modal) gave good adult returns. However, the size difference between the upper and lower modal classes is not an absolute factor in determining marine survival. Growth rate is important as well. Temperature and photoperiod can be manipulated to accelerate growth rates producing more upper mode fish and therefore an increase in 1+ smolt production (Saunders et al. 1994).

Time of release at a given smolt size also has a profound effect on marine survival. Bilton (1978) first investigated this phenomenon by releasing coho smolts in several size categories over a period of 4 months. The optimal smolt size varied with time of release. The maximum return of adult coho salmon resulted from the release of 20-gram smolts just prior to the summer solstice. Very large smolts were associated with more precocious males (jacks) in the population and thus a reduced total adult biomass.

Dramatically increased gill Na^+K^+-stimulated ATPase activity has long been associated with the parr–smolt transformation in salmonids. It is currently being widely used as a method to assess migration readiness, hypoosmoregulatory capability, and potential for good ocean survival of hatchery produced anadromous fishes. In coho salmon, for example, ATPase activity in freshwater normally begins to increase in late April, peaks in late May, and then begins to decline if parr-reversion occurs. If the smolts are transferred into saltwater, the gill ATPase activity will continue to increase. After several more weeks, it usually stabilizes at a level three- to fourfold higher than its initial freshwater activity, indicating that full osmoregulatory capability has been reached (Bjerknes et al. 1992). Larger smolts do not necessarily have the highest gill ATPase activities, although size per se is quite important in seawater tolerance. For example, nonanadromous rainbow trout of >150 grams will continue to develop normally after transfer into 30 ppt seawater with no change in their gill ATPase activity levels.

ATPase development representative of what can be expected during smolting of coho and spring chinook salmon and steelhead trout was discussed in Chapter 2. However, several fish cultural, environmental, and physiological factors must be taken into account before using gill ATPase enzyme activity rates to decide at what time, age, and size smolts should be transferred into seawater net pens for salmon farming operations or released from conservation hatcheries to supplement natural populations. First, the act of migration itself may be required to stimulate gill ATPase activity in some races of anadromous salmonids. Second, certain salmonid stocks have inherently lower gill ATPase enzyme activity levels than others. Finally, as mentioned earlier, crowding, pre-release disease treatments, or low-level contaminant exposure during rearing may have partially or completely inactivated the gill ATPase system. Thus, the pattern of ATPase development can also be used in a positive way, as a sensitive biological method of detecting adverse environmental conditions during freshwater rearing.

Physiological Assessment of Smolt Development

The parr–smolt transformation that enables juvenile anadromous salmonids to migrate to the ocean and continue to grow and develop normally requires a variety of morphological, physiological, biochemical, and behavioral changes. Those of particular interest in assessing stage of smolt development in intensive fish culture operations include: decrease in condition factor, increase in gill ATPase enzyme activity, increased hypoosmoregulatory ability, and changes in the blood concentrations of several hormones. The latter[13] include decreased plasma prolactin concentrations and increased plasma thyroid (T_4), growth hormone, cortisol, and insulin levels.

Hypoosmoregulatory ability as a criterion of smolting success has long been determined by ability to survive for 30 days in 30 ppt seawater (Conte and Wagner 1965). As mentioned earlier, however, salinity tolerance fails to distinguish between large parr and true smolts in many salmonid species. For example, even nonsmolting stocks of rainbow trout can thrive in seawater, provided they are of sufficient size. Experience has also shown that incompletely smolted coho and Atlantic salmon can survive for extended periods in 30 ppt seawater before they

---

13. Summarized in Figure 2.8.

begin to suffer retarded growth (stunting). Although inconvenient, challenge with seawater formulated to increase the salinity to 40 ppt is a better indicator of smolt status and capacity for hypoosmoregulation that reveals undeveloped salinity tolerance within 72–96 hours (Saunders et al. 1985).

Challenge with 30 ppt seawater is more practical and is also an effective test of smolt development if salinity tolerance is assessed by determining the plasma sodium level after 24 hours. True smolts suffer only a mild hypernatremia after abrupt transfer to seawater and can regulate plasma Na^+ to <170 meq/L within 24 hours (Blackburn and Clarke 1987). The seawater challenge/blood sodium test is run by simply transferring 10 fish into 30 ppt seawater at their acclimation temperature and taking blood samples for plasma sodium analyses after 24 hours. Alternatively, following the plasma Na^+ profile of a larger group of fish over a 4-day period can also furnish useful information.

Experience has shown that a plasma sodium concentration of less than 170 meq/L after 24 hours in seawater is predictive of long-term ability to grow and develop normally in the ocean for many stocks of Pacific and Atlantic salmon. Test results accumulated from numerous wild and hatchery-reared coastal coho stocks suggest that well-developed smolts with short outmigration routes can regulate their blood sodium to 150–160 meq/L, or even lower (Blackburn and Clarke 1987). In contrast, stocks that have long outmigration routes from headwaters of rivers such as the Columbia River, will not perform as well. In addition, it must be emphasized that the indicated performance range values are valid only if they are obtained during the first 24–48 hours in seawater because some degree of osmoregulatory homeostasis will eventually develop even in parr, if they survive. However, the energy costs of hypoosmoregulation in nonfunctional or poorly functional smolts are excessive and normal growth cannot be maintained. Stunted coho salmon, in particular are able to regulate their plasma sodium concentrations into the normal range and can survive for several months in seawater net pens, but suffer poor growth, parr-reversion, and eventual mortality.

The seawater challenge/blood sodium test can also be used as an indicator of smolt condition and potential for ocean survival of chinook and sockeye salmon and steelhead trout. The degree of hypernatremia that results in seawater-challenged smolts that are fully functional appears to be similar in all four anadromous species. In the case of fall chinook salmon, however, certain races may normally enter estuaries before the smolting process is fully complete.

Physiological Problems after Release

Following hatchery rearing, downstream migrant smolts must again physiologically cope with numerous environmental stress factors in the rivers and estuaries that can adversely affect saltwater adaptation and subsequent survival after seawater entry. These include dam passage problems, reduced stream flows, elevated water temperatures, predation, gas supersaturation, activation of latent infections, gill infestations by *Icthyobodo*, and, where trucking and barging around dams is employed, handling stress and scale loss.

The smolt mortality during seaward emigration resulting from factors such as unfavorable river temperatures and flows, gas supersaturation, dam passage, and predation have been fairly well documented. However, many other physiological problems occur such as scale loss, activation of latent infections due to the stress of seawater entry, and salinity intolerance due to gill parasite infestations. These problems have received less attention and are briefly discussed here.

Scale Loss. Because of factors such as inadequate fish passage facilities and predation on smolts disoriented after passing through turbines, smolt-hauling operations around dams are in progress in many rivers highly developed for hydropower generation. Unfortunately, various degrees of scale loss can occur during smolt collection and holding, and handling and transport operations. Bouck and Smith (1979) first showed that a substantial mortality occurred after 10 days in seawater when coho smolts were experimentally descaled as little as 10% of the body surface. Scale loss on the ventral surface is apparently the most serious. An estimated 10-day TLm of 50% mortality resulted from a 10% scale loss. Dorsal or lateral descaling was less dramatic but still serious. A delayed mortality following descaling of this magnitude has serious implications for the success of smolt-hauling programs involving stocking into seawater or of interpreting high seas tagging results. A 5-day recovery period in freshwater almost completely restores full salinity tolerance. However, the consequences for smolts with scale loss released into rivers would still be serious because of increased susceptibility to *Saprolegnia* infections. Smolts with scale loss released downstream near the ocean would be expected to show an avoidance reaction to high salinities and would tend to remain in the freshwater/estuarine areas longer than normal, thus further increasing exposure to *Saprolegnia*, *Vibrio*, and predation.

Mineral salt additions at low concentrations can be useful in minimizing effects of handling and scale loss on transported smolts. These include NaCl at 3–5 ppt or $CaCl_2$ at 50 mg/L. Salt (NaCl) additions at 10–15 ppt have considerable potential both for stress mitigation and for *Saprolegnia* control, a major cause of delayed smolt mortality when scale loss occurs. Normally, small amounts of potassium are also added if NaCl at 10 ppt or more is used. Alternatively, diluted seawater (5–15 ppt) can also be effective in mitigating the effects of handling stress and scale loss. A potential side effect to consider is that smolts with scale loss can absorb physiologically undesirable amounts of Mg^{++} and K^{+} from seawater or from mineral salt formulations. As little as 10% (dorsal) scale loss can result in life-threatening hyperkalemia and hypermagnesemia in coho salmon smolts. Even if immediate mortalities do not occur, blood electrolyte imbalances of this magnitude are debilitating and can considerably reduce the ability of affected smolts to survive further stress or escape predation (Wedemeyer, unpublished data).

Activation of Latent Infections. A consensus exists among fish disease biologists that latent bacterial or viral infections activated by the stress of migration and entry into saltwater likely are a significant factor in the total ocean mortality that occurs after smolts enter seawater. Chronic infections such as proliferative kidney disease and bacterial kidney disease seem to be particularly important in this regard. Frantsi et al. (1975) was one of the first to document that juvenile Atlantic salmon released with bacterial kidney disease infections are less able to survive acclimation to seawater and that few heavily infected smolts survive to return as adults. Similarly, bacterial kidney disease (BKD) is known to cause high mortalities in chinook salmon transferred into seawater (Bannerd et al. 1985). More recently, the in-river survival of BKD-infected migrating smolts has been shown to be reduced as well (Pascho et al. 1993). Similar results with furunculosis strongly suggest that smolts released with most subclinical bacterial infections are likely to have the infections activated by the stress of acclimation to saltwater and a delayed mortality will occur after seawater entry. With chronic infections, the majority of the mortalities may not occur for several months. Vibriosis, a stress-mediated bacterial disease, is contracted after seawater entry, both in estuaries and in saltwater net pens. Species released from conservation hatcheries such as chum, pink, or chinook salmon whose life histories include a significant amount of estuarine rearing are especially vulnerable. Acute infections like vibriosis and furunculosis that can also be transmitted in seawater can reach epidemic proportions in the crowded confines of net pen culture within a much shorter time.

Salinity Intolerance Due to Gill Infections. Infections with the fungus *Ichthyophonus* spp. and the freshwater cestode *Eubothrium salvelini* have both been shown to impair the ability of affected smolts to survive seawater entry (Boyce and Clarke 1983; Uno 1990). Infestations of the gill parasite *Ichthyobodo* (*Costia*) can be a particularly important cause of poor survival when smolts are transferred into seawater. Fish cultural experience with coho and chinook smolts suggests that only 15–30 *Ichthyobodo necator* per gill arch will physically damage the chloride cells and cause sufficient epithelial hyperplasia to result in impaired seawater survival. The seawater mortality has exceeded 60% when *Ichthyobodo*-infested juvenile chum salmon are transferred into full strength seawater (Urawa 1993). Prior acclimation at 10 ppt did not improve seawater survival. As previously discussed, an adequate freshwater recovery period should always be allowed after smolts are given formalin treatments.

REFERENCES

Amend, D. F., T. R. Croy, B. A. Goven, K. A. Johnson, and D. H. McCarthy. 1982. Transportation of fish in closed systems: methods to control ammonia, carbon dioxide, pH, and bacterial growth. Transactions of the American Fisheries Society 111:603–611.

Backman, T. W. H., and R. M. Ross. 1990. Comparison of three techniques for the capture and transport of impounded subyearling American shad. Progressive Fish-Culturist 52:246–252.

Banks, J. L. 1994. Raceway density and water flow as factors affecting spring chinook salmon during rearing and after release. Progressive Fish-Culturist 54:137–217.

Banks, J. L., and L. G. Fowler. 1982. Transportation, storage, and handling studies of fall chinook salmon gametes and newly fertilized eggs. Technology Transfer Series No. 82–4. U.S. Fish and Wildlife Service, Abernathy Salmon Cultural Development Center, Longview, Washington.

Banks, J. L., W. G. Taylor, and S. L. Leek. 1979. Carrying capacity recommendations for Olympia area national fish hatcheries. Abernathy Hatchery Technology Development Center, U.S. Fish and Wildlife Service, Washington, D.C.

Banner, C. R., J. S. Rhovec, and J. L. Fryer. 1985. *Renibacterium salmoninarum* as a cause of mortality among chinook salmon in saltwater. Journal of the World Mariculture Society 14:236–239.

Barton, B. B., and G. K. Iwama. 1991. Physiological changes in fish from stress in aquaculture with emphasis on the response and effects of corticosteroids. Annual Review of Fish Diseases 1:3–26.

Basu, S. P. 1959. Active respiration of fish in relation to ambient concentrations of oxygen and carbon dioxide. Journal of the Fisheries Research Board of Canada 16:175–212.

Bilton, H. T. 1978. Returns of adult coho salmon in relation to mean size and time at release of juveniles. Canadian Fisheries and Marine Service Technical Report 832. Department of Fisheries and Oceans, Ottawa, Canada.

Bjerknes, V., J. Duston, D. Kness, and P. Harmon. 1992. Importance of body size for acclimation of underyearling Atlantic salmon parr (*Salmo salar*) to seawater. Aquaculture 104:357–366.

Blackburn, J., and W. C. Clarke. 1987. Revised procedure for the 24-hour seawater challenge test to measure seawater adaptability of juvenile salmonids. Canadian Technical Report in Fisheries and Aquatic Sciences 1515.

Blackburn, J., and W. C. Clarke. 1990. Lack of density effect on growth and smolt quality in zero-age coho salmon. Aquacultural Engineering 9:121–130.

Blazer, V. S. 1992. Nutrition and disease resistance in fish. Annual Review of Fish Diseases 1:309–323.

Bouck, G. R., and D. A. Johnson. 1979. Medication inhibits tolerance to seawater in coho salmon smolts. Transactions of the American Fisheries Society 108:63–66.

Bouck, G. R., and S. D. Smith. 1979. Mortality of experimentally descaled smolts of coho salmon (*Oncorhynchus kisutch)* in fresh and saltwater. Transactions of the American Fisheries Society 108:67–69.

Boyce, N. P., and W. C. Clarke. 1983. *Eubotherium salvelini* (Cestoda: *Pseudophyllidea*) impairs seawater adaptation of migrant sockeye salmon yearlings (*Oncorhynchus nerka*) from Babine Lake, British Columbia. Canadian Journal of Fisheries and Aquatic sciences 40:821–824.

Brannon, E., C. Feldman, and L. Donaldson. 1982. University of Washington zero-age coho salmon smolt production. Aquaculture 28:195–200

Bullock, G., R. Herman, J. Heinen, A. Noble, A. Weber, and J. Hankins. 1994. Observations on the occurrence of bacterial gill disease and amoeba gill infestation in rainbow trout cultured in a water recirculation system. Journal of Aquatic Animal Health 6:310–317.

Buss, K. 1981. An approach to functional, economical, and practical fish culture through better bioengineering. Pages 227–234 *in* L. Allen and E. Kinney (eds.), Bioengineering Symposium for Fish Culture, American Fisheries Society, Bethesda Maryland.

Caldwell, C. A., and J. Hinshaw. 1995. Tolerance of rainbow trout to dissolved oxygen supplementation and a *Yersinia ruckeri* challenge. Journal of Aquatic Animal Health 7:168–171.

Carmichael, G. J., and J. R. Tomasso. 1988. Survey of fish transportation equipment and techniques. Progressive Fish-Culturist 50:155–159.

Carmichael, G. J., J. R. Tomasso, B. A. Simco, and K. B. Davis. 1984. Characterization and alleviation of stress associated with hauling largemouth bass. Transactions of the American Fisheries Society 113:778–785.

Ceballos, J. R., S. W. Petit, and J. L. Mckern. 1991. Fish transportation oversight team annual report—FY 1990. Transport operations on the Snake and Columbia Rivers. NOAA Technical Memorandum NMFS F/NWR-29. U.S. Department of Commerce, National Oceanic and Atmospheric Administration, National Marine Fisheries Service, Washington, D.C.

Clarke, W. C., and J. E. Shelbourne. 1986. Delayed photoperiod produces more uniform growth and greater seawater adaptability in underyearling coho salmon (*Oncorhynchus kisutch*). Aquaculture 56:287–299.

Clarke, W. C., J. E. Shelbourne, and J. R. Brett. 1981. Effect of artificial photoperiod cycles, temperature, and salinity on growth and smolting in underyearling coho (*Oncorhynchus kisutch*), chinook (*O. tshawytscha*), and sockeye (*O. nerka*) salmon. Aquaculture 22: 105–116.

Colt, J. E., and G. Tchobanoglous. 1981. Design of aeration systems for aquaculture. Pages 138–148 *in* L. Allen and E. Kinney (eds.), Bioengineering Symposium for Fish Culture, American Fisheries Society, Bethesda Maryland.

Colt, J., K. Orwicz, and G. L. Bouck. 1991. Water quality considerations and criteria for high density fish culture with supplemental oxygen. American Fisheries Society Symposium 10:372–385.

Conte, F. P., and H. H. Wagner. 1965. Development of osmotic and ionic regulation in juvenile steelhead trout *Salmo gairdneri*. Comparative Biochemistry and Physiology 14:603–620.

Davis, K. B., N. C. Parker, and M. A. Suttle. 1982. Plasma corticosteroids and chlorides in striped bass exposed to tricaine methane sulfonate, quinaldine, etomidate, and salt. Progressive Fish-Culturist 44:205–210.

Ewing, R. D., and S. K. Ewing. 1995. Review of the effects of rearing density on survival to adulthood for Pacific salmon. Progressive Fish-Culturist 57:1–25.

Fowler, L. G. 1980. Substitution of soybean and cottonseed products for fish meal in diets fed to chinook and coho salmon. Progressive Fish-Culturist 42:87–91.

Frantsi, C., T. C. Flewelling, and K. G. Tidswell. 1975. Investigations on corynebacterial kidney disease and *Diplostomulum sp*. (eye fluke) at Margaree hatchery. Technical Report MAR-T-75–9. New Brunswick, N.S., Canada.

Fraser, J. M., and F. W. H. Beamish. 1969. Blood lactic acid concentrations in brook trout, *Salvelinus fontinalis*, planted by air drop. Transactions of the American Fisheries Society 98:263–267.

Fredericks, K. T., W. H. Gingrich, and D. C. Fater. 1993. Comparative cardiovascular effects of four fishery anesthetics in rainbow trout (*Oncorhynchus mykiss*). Comparative Biochemistry and Physiology 104C:477–483.

Ghittino, P. 1989. Nutrition and fish diseases. Pages 681–713 *in* J. Halver (ed.), Fish Nutrition, 2nd ed. Academic Press, New York.

Halver, J. (editor). 1989. Fish Nutrition, 2nd ed. Academic Press, New York.

Hardy, R. W., and C. V. Sullivan. 1983. Canola meal in rainbow trout (*Salmo gairdneri*) production diets. Canadian Journal of Fisheries and Aquatic Sciences 40:281–286.

Haskell, D. C. 1941. An investigation on the use of oxygen in transporting trout. Transactions of the American Fisheries Society 70:149–160.

Haswell, M. S., G. J. Thorpe, L. E. Harris, T. C. Mandis, and R. E. Rauch. 1982. Millimolar quantities of sodium salts used as prophylaxis during fish hauling. Progressive Fish-Culturist 44:179–182.

Hauck, H. K. 1986. Gas bubble disease due to helicopter transport of young pink salmon. Transactions of the American Fisheries Society 115: 630–635.

Hendricks, J. D., and G. S. Baily. 1989. Adventitious toxins. Pages 606–645 *in* J. Halver (ed.), Fish Nutrition, 2nd ed. Academic Press, New York.

Hendricks, J. D., R. O. Sinnhuber, M. C. Henderson, and D. R. Buhler. 1981. Liver and kidney pathology in rainbow trout (*Salmo gairdneri*) exposed to dietary pyrrdizidine (*Senecio*) alkaloids. Experimental Molecular Pathology 35:170–183.

Hendricks, J. D., T. R. Meyers, D. W. Shelton, J. L. Casteel, and G. S. Bailey. 1985. Hepatocarcinogenicity of benzo [*a*] pyrene to rainbow trout by dietary exposure and intraperitoneal injection. Journal of the National Cancer Institute 74:839.

Hickling, C. F. 1971. Fish Culture. Faber and Faber, London.

Higgs, D. A., J. R. McBride, J. R. Markert, B. Dosanjh, M. D. Plotnikoff, and W. C. Clarke. 1982. Evaluation of Tower and Candle rapeseed (canola) meal and Bronowski rapeseed protein concentrate as protein supplement in practical dry diets for juvenile chinook salmon (Oncorhynchus tshawytscha). Aquaculture 29:1–31.

Hobe, H., C. M. Wood, and M. G. Wheatly. 1984. The mechanisms of acid-base and ionregulation in the freshwater rainbow trout during environmental hyperoxia and subsequent normoxia. 1. Extra- and intracellular acid-base status. Respiratory Physiology 55:139–154.

Hodson, P. V., B. R. Blunt, and D. J. Spry. 1978. Chronic toxicity of water-borne and dietary lead to rainbow trout (*Salmo gairdneri*) in Lake Ontario water. Water Research 12:869–878.

IDC (Illinois Department of Conservation). 1979. Biological criteria: fish hatchery program State of Illinois. Illinois Department of Conservation, Project Report 102–010–006. Krammer, Chin, and Mayo Inc., Seattle, Washington.

Johnson, S. K. 1979. Transport of live fish. Publication FDDL-F14, Fish Disease Diagnostic Laboratory. Texas Agricultural Extension Service, College Station, Texas.

Kebus, M. J., M. T. Collins, M. S. Brownfield, C. H. Amundson, T. B. Kayes, and J. A. Malison. 1992. Effects of rearing density on the stress response and growth of rainbow trout. Journal of Aquatic Animal Health 4:1–6.

Ketola, H. G., C. E. Smith, and G. A. Kindschi. 1989. Influence of diet and oxidative rancidity on fry of Atlantic and coho salmon. Aquaculture 79:417–423.

Klontz, G. W. 1992. Environmental requirements and environmental diseases of salmonids. Pages 33–34 *in* M. Stoskopf (ed.), Fish Medicine. W. B. Saunders, Philadelphia, Pennsylvania.

Kreiberg, H., and J. Powell. 1991. Metomidate sedation reduces handling stress in chinook salmon. World Aquaculture 22:58–59.

Larson, D. 1984. A new mechanism designed for helicopter fry releases. Page 85 *in* D. F. Alderdice, F. E. A. Wood, and D. W. Narver (eds.), Salmonid enhancement program—preliminary notes on new information in salmonid hatchery propagation. Canadian Data Report Fisheries and Aquatic Science 496. Department of Fisheries and Oceans, Vancouver, B.C., Canada.

Leatherland, K. F., and R. Sonstegard. 1984. Pathobiological response of feral teleosts to environmental stressors: interlake studies of the physiology of Great Lakes salmon. Pages 116–140 *in* V. Cairns, P. Hodson and J. Nriagu (eds.), Contaminant Effects on Fisheries. Wiley, New York.

Lewis, M. A., T. R. Walter, and R. D. Ewing. 1993. Evaluation of inventory procedures for hatchery fish. 2. Variation in specific gravities of Pacific salmonids during rearing. Progressive Fish-Culturist 56:160–168.

Liao, P. 1971. Water requirements of salmonids. Progressive Fish-Culturist 38:210–218.

Maule, A. G., R. A. Tripp, S. L. Kaattari, and C. D. Schreck. 1989. Stress alters immune function and disease resistance in chinook salmon (*Oncorhynchus tshawytscha*). Journal of Endocrinology 120:135–142.

Mazik, P. M., B. A. Simco, and N. C. Parker. 1991. Influence of water hardness and salts on survival and physiological characteristics of striped bass during and after transport. Transactions of the American Fisheries Society 120:121–126.

Mazur, C. F., and G. K. Iwama. 1993. Handling and crowding stress reduces the number of plaque forming cells in Atlantic salmon. Journal of Aquatic Animal Health 5:98–101.

McCraren, J. P. 1978. History. Pages 1–6 *in* C. Smith (ed.), Manual of Fish Culture, Section G: Fish Transportation. U.S. Fish and Wildlife Service, Washington, D.C.

McCraren, J. P., and R. M. Jones. 1978. Suggested approach to computing and reporting loading densities for fish transport units. Progressive Fish–Culturist 40:169.

Meade, J. W. 1991. Application of the production capacity assessment bioassay. American Fisheries Society Symposium 10:365–367.

Meehan, W. R., and L. Revet. 1962. The effect of tricaine methane sulfonate (MS-222) and/or chilled water on oxygen consumption of sockeye salmon fry. Progressive Fish-Culturist 24:185–187.

Pascho, R. P., E. Elliott, and S. Achord. 1993. Monitoring of the in-river migration of smolts from two groups of spring chinook salmon, *Oncorhynchus tshawytscha* (Walbum), with different profiles of *Renibacterium salmoninarum* infection. Broodstock Management and Egg and Larval Quality. Aquaculture and Fisheries Management 24:163–169.

Pennell, W. 1991. Fish Transportation Handbook. Province of British Columbia, Ministry of Fisheries. Victoria, B.C., Canada.

Peterson, H. H. 1973. Adult returns to date from hatchery-reared one-year-old smolts. Pages 219–226 *in* M. V. Smith and W. M. Carter (eds.), International Atlantic Salmon Symposium, vol. 4. International Atlantic Salmon Foundation, New York.

Phillips, A. M., and D. R. Brockway. 1954. Effect of starvation, water temperature and sodium amytal on the metabolic rate of brook trout. Progressive Fish-Culturist 16 (2):65–68.

Piper, R. G., I. B. McElwain, L. E. Orme, J. P. McCraren, L. G. Fowler, and J. R. Leonard. 1982. Fish Hatchery Management. United States Department of the Interior, Fish and Wildlife Service, Washington, D.C.

Ray, L. 1981. Channel catfish production in geothermal water. Pages 192–195 *in* L. Allen and E. Kinney (eds.), Proceedings of the Bioengineering Symposium for Fish Culture, American Fisheries Society, Bethesda, Maryland.

Richardson, N. L., D. A. Higgs, R. M. Beames, and J. R. McBride. 1985. Influence of dietary calcium, phosphorous, zinc, and sodium phytate level on cataract incidence, growth, and histopathology in juvenile chinook salmon (*Oncorhynchus tshawytscha*). Journal of Nutrition 115:553–567.

Roberts, R. J., and A. M. Bullock. 1989. Nutritional pathology. Pages 424–475 *in* J. Halver (ed.), Fish Nutrition, 2nd ed. Academic Press, New York.

Robinson, E. H., S. D. Rawles, and R. R. Stickney. 1984. Evaluation of glanded and glandless cottonseed products in catfish diets. Progressive Fish Culturist 46:92–97.

Saunders, R. L., E. B. Henderson, and P. R. Harmon. 1985. Effects of photoperiod on juvenile growth and smolting of Atlantic salmon and subsequent survival and growth in seacages. Aquaculture 45:55–65.

Saunders, R. L., J. Duston, and T. J. Benfey. 1994. Environmental and biological factors affecting growth dynamics in relation to smolting of Atlantic salmon, *Salmo salar* L. Aquaculture and Fisheries Management 25:9–20.

Saxton, A. M., R. N. Iwamoto, and W. K. Hershberger. 1983. Smoltification in the net-pen culture of accelerated coho salmon, *Oncorhynchus kisutch* Walbaum: prediction of saltwater performance. Journal of Fish Biology 22:363–370.

Schreck, C. B., R. Patinõ, C. Pring, J. Winton, and J. Holway. 1985. Effects of rearing density on indices of smoltification and performance of coho salmon, *Oncorhynchus kisutch*. Aquaculture 45:345–358.

Shepherd, B. G., and G. F. Bérézay. 1987. Fish transport techniques in common use at salmonid enhancement facilities in British Columbia. Canadian Manuscript Report of Fisheries and Aquatic Sciences No. 1946. Fisheries and Oceans Canada, Quebec.

Skak, C., and E. Baatrup. 1993. Quantitative and histochemical demonstration of mercury deposits in the inner ear of trout, *Salmon trutta*, exposed to dietary methylmercury. Aquatic Toxicology 25:55–70.

Smith, C. E., 1978. Transportation of salmonid fishes. Pages 9–41 *in* C. Smith (ed.), Manual of Fish Culture, Section G: Fish Transportation. U.S. Fish and Wildlife Service, Washington, D.C.

Smith, R. R. 1977. Studies on the energy metabolism of cultured fishes. Doctoral Dissertation, Graduate School of Cornell University, Cortland, New York.

Smith, S. D., R. W. Gould, W. S. Zaugg, L. W. Harrell, and C. W. Mahnken. 1987. Safe prerelease disease treatment with formalin for fall chinook salmon smolts. Progressive Fish-Culturist 49:96–99.

Smith, S. D., R. W. Gould, W. S. Zaugg, L. W. Harrell, and C. V. Mahnken. 1995. Prerelease disease treatment with potassium permanganate for fall chinook salmon smolts. Progressive Fish-Culturist 57:102–106.

Soderberg, R. W. 1995. Flowing Water Fish Culture. Lewis Publishers, Boca Raton, Florida.

Soderberg, R. W., J. W. Meade, and L. A. Redell. 1993 Growth, survival, and food conversion of Atlantic salmon reared at four different densities with common water quality. Progressive Fish-Culturist 55:29–31.

Sower, S. A., and R. S. Fawcett. 1991. Changes in gill Na^+,K^+-ATPase, thyroxine, and triiodothyronine of coho salmon held in two different rearing densities during smoltification. Comparative Biochemistry and Physiology 99A:85–89.

Spinelli, J., C. R. Houle, and J. C. Wekell. 1983. The effect of phytates on the growth of rainbow trout (*Salmo gairdneri*) fed purified diets containing varying quantities of calcium and magnesium. Aquaculture 30:71–84.

Stickney, R. R. 1983. Care and handling of live fish. Pages 85–93 *in* L. A. Nielson and D. L. Johnson (eds.), Fisheries Techniques. American Fisheries Society, Bethesda, Maryland.

Summerfelt, R. C., and L. S. Smith. 1990. Anesthesia, surgery, and related techniques. Pages 213–272 *in* C. Schreck and P. Moyle (eds.), Methods for Fish Biology. American Fisheries Society, Bethesda, Maryland.

Thrush, M. A., N. J. Duncan, and N. R. Bromage. 1994. The use of photoperiod in the production of out-of-season Atlantic salmon (Salmo salar) smolts. Aquaculture 121:29–44.

Tucker, C. S., and E. H. Robinson. 1990. Channel Catfish Farming Handbook. Van Nostrand Reinhold, New York.

Ubels, J. H., and H. F. Edelhauser. 1987. Effects of corneal epithelial abrasion on corneal transparency, aqueous humor composition, and lens of fish. Progressive Fish-Culturist 49:219–224.

Uno, M. 1990. Effects of seawater acclimation on juvenile salmonids infected with *Tetraonchus* (Monogenea) and *Ichthyophonus* (Phycomycetes). Fish Pathology 24:15–19.

Urawa, S. 1993. Effects of *Ichthyobodo necator* infections on seawater survival of juvenile chum salmon (*Oncorhynchus keta*). Aquaculture 110:101–110.

Voss, S. D., D. W. Shelton, and J. D. Hendricks. 1982. Effects of dietary Aroclor 1254 and cyclopropene fatty acids on hepatic enzymes in rainbow trout. Archives of Environmental Contamination and Toxicology 11:87–91.

Wedemeyer, G. A. 1982. Effects of environmental stressors in aquacultural systems on quality, smoltification and early marine survival of anadromous fish. Pages 155–170 *in* B. Melteff and R. Neve (eds.), Proceedings of the North Pacific Aquaculture Symposium. Sea Grant Report 82–2, University of Alaska, Fairbanks.

Wedemeyer, G. A. 1996. Handling and transportation of salmonids. Pages 00–00 *in* W. Pennel and B. Barton (eds.), Developments in Aquaculture and Fisheries Management. Elsevier, Holland.

Wedemeyer, G. A., and C. P. Goodyear. 1984. Diseases caused by environmental stressors. Pages 424–434 *in* O. Kinne (ed.), Diseases of Marine Animals, vol. 4, part 1: Pisces. Biologische Anstalt Helgoland, Hamburg, FRG.

Wedemeyer, G. A., R. L. Saunders, and W. C. Clarke. 1980. Environmental factors affecting smoltification and early marine survival of anadromous salmonids. U.S. National Marine Fisheries Service Marine Fisheries Review 42(6):1–14.

Westers, H. 1984. Principles of Intensive Fish Culture (a manual for Michigan's state fish hatcheries). Michigan Department of Natural Resources, Lansing, Michigan.

Wilson, R. P., and W. E. Poe. 1985. Effects of feeding soybean meal with varying trypsin inhibitor activities on growth of fingerling channel catfish. Aquaculture 46:19–25.

Wobeser, G. 1975. Prolonged oral administration of methyl mercury chloride to rainbow trout (*Salmo gairdneri*) fingerlings. Journal of the Fisheries Research Board of Canada 32:2015.

Woodward, B., L. G. Young, and A. K. Lun. 1983. Vomitoxin in diets for rainbow trout. Aquaculture 35:93–101.

Woodward, D. F., W. G. Brumbaugh, A. J. DeLonay, E. E. Little, and C. E. Smith. 1993. Effects on rainbow trout fry of a metals contaminated diet of benthic invertebrates from the Clark Fork River, Montana. Transactions of the American Fisheries Society 123:51–62.

Youngs, W. D., and M. B. Timmons. 1991. Engineering aspects of intensive aquaculture. Proceedings of the Aquaculture Symposium, Cornell University, Ithica, New York.

5
Biological Interactions

INTRODUCTION

In addition to the interactions between fish and the chemical and physical factors discussed in previous chapters, fish also continually interact with each other and with the microorganisms present in the rearing environment. These biological interactions must likewise be carefully managed if health and physiological quality are to be maintained.

The fish–fish interactions of interest in aquaculture are behavioral and include aggression, dominance hierarchies, and interspecific competition. In natural populations, behavioral interactions are adaptive and enhance access to resources such as food, space, favorable water quality, and breeding partners. Under hatchery conditions, however, these behaviors usually do little to provide such resources and may instead become a source of pathology.

The biological interactions between fish and the microorganisms present in the rearing environment are potentially a much more serious source of health and quality problems. Fish–microorganism interactions occur at the physiological/biochemical level and are strongly modified by environmental conditions. Under favorable physicochemical conditions, these interactions are most often harmless or beneficial. Under unfavorable physicochemical conditions, serious disease problems can occur and the resulting economic losses can be catastrophic.

INTERACTIONS BETWEEN FISH

As mentioned, a number of behaviors have evolved in the interactions between fish that enhance individual success in the endless competition for food and other resources such as space, favorable water quality, cover for protection against predators, and breeding partners. Under natural conditions, these behaviors are adaptive and serve to regulate life in groups. The behaviors of interest are agonistic[14] and are used to intimidate rivals and hinder their access to particular resources. Aggression is used to establish individual fish as territory holders, as dominants in intraspecific hierarchical social systems, and to enhance success in interspecific competition. Some aggressive behaviors, such as biting, charging, chasing, and ramming, can inflict serious physical injuries. Others, such as lateral displays, intimidate rivals without requiring direct physical contact. During the initial establishment of territories or social hierarchies, aggressive interactions tend to be frequent and intense. After social systems have been established, however, conflicts usually become less frequent and relatively stable relationships may develop between dominants and subordinates. Residency within established territories often becomes long term. As an alternative to permanent submission, less successful competitors usually also have the option of moving to another location. For example, fish can often find another feeding territory in an area where competition is less and frequent conflicts with dominants can be avoided. Of course, food availability may be lower and risk of predation higher.

When fish are reared under production hatchery conditions, such as in raceways with broadcast feeders, behaviors such as aggression do little to provide food or cover and simply become aberrant, causing unnecessary stress and physical injury. Migration as an alternative to frequent aggressive conflicts with dominants is usually not an option. Broken or damaged fins from repeated biting, and reduced growth due to chronic stress in the defeated individuals are common sequelae. Other behavioral interactions that serve little purpose under intensive culture conditions and can adversely affect health and physiological condition include interspecific competition, social dominance hierarchies, and territoriality. Interspecific competition does not usually have to be considered in intensive culture operations but can be a source of pathology in extensive polyculture systems or in experimentally mixed populations. For example, when brook trout were held with equal numbers of brown trout in artificial streams, they suffered a 33% mortality from *Saprolegnia* infections. Brown trout were never infected nor were brook trout when they were held alone at the same density (Wald and Wilzbach 1992).

Health problems resulting from social dominance in fish–fish interactions are a relatively common occurrence in aquaculture systems. Dominance hierarchies are based largely on fish size and can form relatively quickly in many species, particularly at low rearing densities. Aggression is the prevalent method used to establish and maintain hierarchies. Fin nipping, scale loss from ramming, reduced growth, pathological changes in gastrointestinal tissue, and increased susceptibility to infectious diseases due to chronic physiological stress can all occur in de-

14. Aggressive or defensive interactions between individuals such as fighting, fleeing, or submitting.

feated individuals. In eel culture, for example, subordinated fish in the dominance hierarchies that form show chronically elevated levels of corticosteroid (stress) hormones, atrophy of the gastric mucosa, and a suppressed immune protection system (Ejike and Schreck 1980; Peters and Hong 1984). In tilapia culture, leucocyte function is impaired in defeated fish (Cooper et al. 1989).

The fact that dominated fish also grow slower reinforces the size-based hierarchies and badly skewed or bimodal size distributions can develop in the ponds in a matter of weeks. The slower growth of subordinated fish in social hierarchies is due to a combination of the reduced access to feed, the caloric energy costs of chronic stress, and the anorexia that occurs in chronically stressed animals in general. For example, subordinates in juvenile steelhead trout populations grow significantly slower than behavioral dominants even though food intake is experimentally equalized. Subordinates are also less active; presumably also in response to the caloric energy costs of the chronic stress (Abbot and Dill 1989). The rationale for periodic size grading in hatcheries is that larger fish tend to be the dominants and if they are removed, growth of the remaining smaller fish usually improves. In addition, size grading does facilitate feeding because a smaller range of pellet sizes can be used to feed the graded groups. In some hatcheries, periodic size grading is routine because it is assumed to disrupt social hierarchies and thus reduce the divergent growth between dominant and subordinate individuals. However, size grading is not always beneficial. In less domesticated species, a greater degree of size variation may occur naturally because of their greater degree of genetic variability. In addition, although stream dwelling species such as salmonids do show territorial and agonistic behavior under low density (natural) conditions, it is much less prevalent under the high rearing densities used in intensive culture. In the schooling behavior that develops at high raceway densities, imitation rather than aggression is characteristic. This, together with the frequent or automatic feeding typical of intensive culture, tends to reduce size variation independently of grading (Wallace and Kolbeinshavn 1988). In salmon hatcheries producing fish for conservation, size grading is sometimes used as a release strategy. As discussed in Chapter 4, Atlantic salmon are somewhat of a special case because of the bimodal size distribution that normally develops in this species. The juveniles must be graded in the fall to separate potential smolts (upper mode, >12 cm) from the smaller lower mode parr regardless of whether the fish are to be graded later in the spring or not.

In addition to the use of aggression to form social systems, hatchery fish will also become aggressive if they are allowed the opportunity to compete for food. For example, juvenile Pacific salmon normally behave as loosely schooling fish under raceway rearing conditions. However, if they are fed from a point source (e.g., a single demand feeder), some individuals will begin to show aggressive rather than schooling behavior and begin intimidating competitors and monopolizing food. As in dominance hierarchies, ragged or broken fins from repeated biting, and slower growth from stress-induced anorexia and reduced access to food will occur in the defeated individuals.

The challenge for fish culture is to identify and provide rearing conditions under which behaviors such as aggression or social dominance are no longer effective in acquiring space, favorable water quality, food, or other resources. Theoretically, fish will only use aggressive behavior to obtain or defend food if its density and distribution, and the density and distribution of competitors, make aggression bioenergetically more profitable than using alternative tactics such as scrambling (Grant 1993). Thus, practical experience shows that the skewed size distributions resulting from demand feeding can be minimized by increasing the number of feeding stations per raceway, or by increasing the amount of feed released per trigger strike (or by using broadcast feeders). Under these conditions, it becomes easier for subordinates to take food not eaten by dominants or to simply wait until the dominants have fed to satiation and have dispersed. Thus, the energy cost of defeating other fish in order to monopolize food becomes greater than the energy obtained from the feed and aggressive behavior tends to decline. Aggressive behavior can also be modified by making adjustments to rearing densities. In tilapia, an important food fish which is aggressive, practical experience has shown that both aggressive and sexual behavior are density dependent and can be shifted toward the more desirable schooling behavior simply by increasing fish production levels. This is not the case with all species, however, as shown by the fact that smaller eels may fail to feed if larger eels are present even after the larger eels are satiated and food is still available (Knights 1987). In most salmonids, agonistic behavior does tend to decrease at higher fish densities as does the establishment of dominant/subordinate hierarchies. However, even though the number of agonistic interactions initiated per individual may decline as the fish density is increased, total aggression in the raceway/pond may remain high as evidenced by the persistence of damaged fins, particularly in such species as steelhead. Nonetheless, behavioral considerations probably offer significant opportunities for health improvements in the production of most species by intensive culture methods.

It is also important to consider the evolved behavioral tendencies of the species to be reared when an intensive culture system is initially designed. The objective is to avoid imposing rearing conditions that inadvertently stimulate aggressive behavior or that exacerbate the effects of normal social conflicts. For example, rearing a lake-dwelling (schooling) species at densities far greater than ever found in nature may present few problems. However, rearing a stream-dwelling territorial species at densities higher than those found in nature may provoke increased fin nipping and other aggressive acts by forcing continuous encounters between individuals. It is true that most fishes can adapt aspects of their behavior to a range of environmental conditions. However, forcing a strongly territorial species to adopt nonterritorial, schooling behavior by raising the fish density so high that it precludes individual monopolization of available food may simply cause other types of problems. Hatchery designs which incorporate no visual barriers to hinder direct eye contact between dominants and subordinates may also inadvertently stimulate aggressive behavior in some species. Feeding practices that stimulate aggression by delivering small amounts of food to small predictable areas have already been discussed.

The behavior modification that can occur in fish reared under intensive conditions may be undesirable if the fish are intended to supplement wild populations after release. For example, hatchery-reared cutthroat trout that are transferred to artificial streams seem to be more aggressive than their wild conspecifics, as evidenced by their greater use of fast flowing water and a higher incidence of lateral displays and biting. Hatchery-reared coho salmon also seem to be more aggressive than wild juveniles (Swain and Riddle 1990). These behaviors appear to confer little survival advantage and the energy expenditure for the unnecessary aggression may be one factor that reduces efficiency in the use of hatchery fish to supplement wild runs (Mesa 1991). In contrast, when Atlantic salmon fry from hatchery stocks are released into artificial streams to compete for territories with wild fry, they are generally less aggressive. This may be because there is less selection pressure for dominance behavior in the rearing environments used in the intensive culture of these fish (Norman 1987). Acclimation to the frequent episodes of stress (size grading, pond cleaning) encountered during hatchery rearing also seems to attenuate the natural physiological response to stressful challenges. If hatchery coho and chinook salmon are stocked as fry and allowed to acclimate to a natural stream environment for several months, they show the same heightened cortisol response to stressful challenges as wild smolts do (Salonius and Iwama 1993).

INTERACTIONS BETWEEN FISH AND MICROORGANISMS

The interactions between fish and microorganisms that are present in the aquatic environment are a potentially serious source of mortality. These interactions occur at the physiological/biochemical level and are strongly affected by physicochemical conditions in the rearing environment. Fish–microorganism interactions are often harmless or even beneficial. However, some of the viruses, bacteria, fungi, and protozoa are pathogenic and their interactions with fish, as affected by water quality factors and the stress of rearing conditions, can result in infections and epizootic diseases. Fortunately, the number of microorganisms that are directly pathogenic is relatively small. Of all the known bacteria for example, species from fewer than 25 genera are responsible for disease problems in fishes cultured in either marine or freshwater systems (Austin and Allen-Austin 1985). Of these, some, such as *Aeromonas hydrophila* (motile *Aeromonas* septicemia), cause epizootics in a wide variety of cultured freshwater fishes including ornamental species. Others, such as *Renibacterium salmoninarum* (bacterial kidney disease), and *Edwardsiella ictaluri* (enteric septicemia) infect only certain species (salmonids and channel catfish respectively). A few, such as *Lactobacillus* spp. (lactic acid producing bacteria) are of limited pathogenicity and are typically only found as secondary invaders in fish experiencing handling, spawning, or other types of stress. However, one *Lactobacillus* species (*L. piscicola*), is potentially a primary pathogen (Austin and Allen-Austin 1987).

Fish disease biologists and other fishery personnel should have a general understanding of the characteristics of fish pathogens because these characteristics can be used in disease control. Broadly speaking, fish pathogens can be classed as being either obligate or facultative. Obligate pathogens require a living host in order to grow and reproduce and cannot obtain the nutrients required to survive and remain infectious suspended in the water column or attached to sediment particles. When shed into the water by an infected host, however, they usually do remain viable long enough to be transmitted horizontally (fish-to-fish). This group of microorganisms includes all the viruses, some protozoan parasites (*Ichthyophthirius multifiliis*, *Chilodonella* and *Ichthyobodo* spp., *Myxobolus cerebralis*), and a few bacteria such as *Aeromonas salmonicida* (furunculosis), *Yersinia ruckeri* (enteric redmouth disease), and *Renibacterium salmoninarum* (bacterial kidney disease). Thus, obligate pathogens will not normally be found free in a hatchery water supply unless aquatic animal life is also present to serve as a reservoir of infection. In turn, diseases caused by this group of microorganisms will not normally occur in hatchery fish unless other fish, amphibians, or eggs infected with obligate pathogens also reside in the water supply or are introduced. Provided that the fish are not already infected with pathogens that can be transmitted vertically, hatcheries using springs or wells (closed water sources) can minimize or even prevent disease problems caused by obligate pathogens by simply removing any resident fish, frogs, or other amphibians that may be present in the water supply. In hatcheries using water from open sources such as small streams, it is sometimes possible to reduce obligate pathogen numbers by using fish screens to keep resident fish that are reservoirs of infection from moving above the hatchery intakes during their spawning migration. In either case, it is important to note that although obligate pathogens cannot live indefinitely after being shed into the water, they may still survive and remain capable of infecting the hatchery fish for a number of days or even a few weeks depending on the water chemistry. Survival curves for the obligate pathogens *A. salmonicida* (furunculosis), and the IHN (infectious hematopoietic necrosis) virus suspended in distilled, soft, or hard lake waters are shown in Figures 5.1 and 5.2 as examples.

Facultative microbial pathogens do not require aquatic animal life forms as hosts and can obtain sufficient nutrients to live and reproduce while attached to

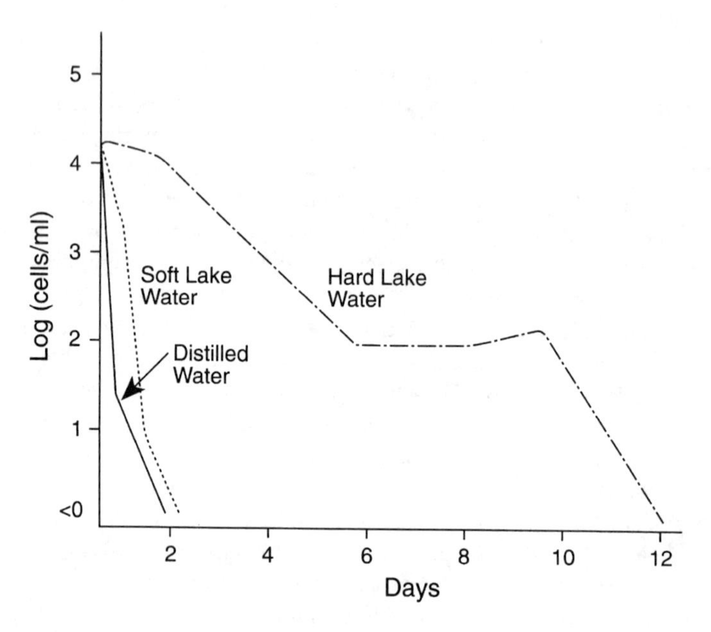

Figure 5.1. Survival curves for the bacterial pathogen *A. salmonicida* (furunculosis) suspended in distilled, soft, or hard lake waters.

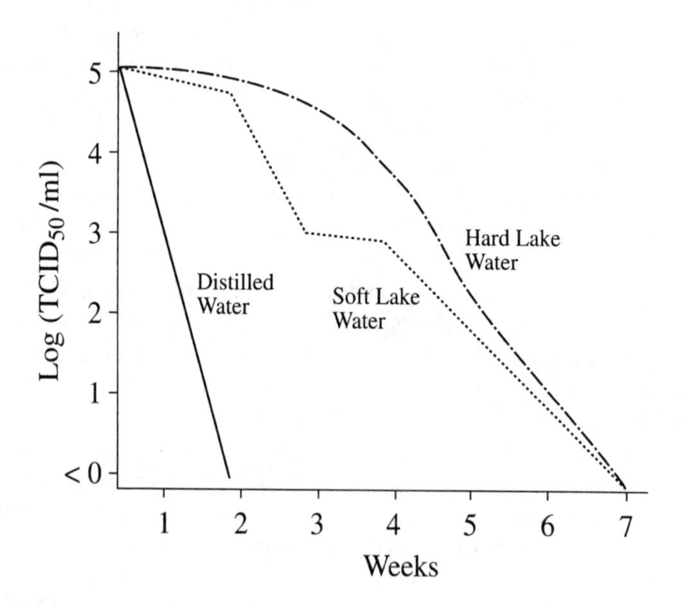

Figure 5.2. Survival curves for the IHN (infectious hematopoietic necrosis) virus suspended in distilled, soft, or hard lake waters.

aquatic plants, suspended organic material (detritus), or even sediment particles. Facultative pathogens are normally ubiquitous in natural waters but cause few observed cases of disease because the low numbers of the pathogens and low population density of most fish populations result in a fish—pathogen–environment relationship favorable to the fish. The low fish density also serves as a biological restraint on the horizontal transmission of pathogens if individual fish do become infected.

In aquaculture, fish disease problems caused by facultative pathogens can be troublesome. Examples in freshwater culture include bacteria such as *Aeromonas hydrophila* (motile *Aeromonas* septicemia), *Flexibacter columnaris* (columnaris disease), *Edwardsiella ictaluri* (enteric septicemia of catfish), many external protozoan parasites (*Trichodina* spp.), and fungi such as *Saprolegnia* spp. In marine net pen culture *Vibrio anguillarum* (vibriosis) is a chronic problem. Since facultative pathogens are normally ubiquitous in surface waters, removing fish or other aquatic life forms will have little effect. Closed water supplies (springs, wells) that are free of aquatic animals and thus, obligate pathogens, are also generally free of facultative pathogens as well because of the extensive filtering of nutrients (carbon and nitrogen sources) by subterranean sands and gravel. Unfortunately, facultative pathogens will usually quickly recolonize the water as soon as it is exposed to the atmosphere in headboxes, tanks, ponds, or raceways.

A general approach to the control of both classes of fish pathogens is water treatment with ultraviolet light (UV), ozone, chlorine, or other disinfectants (discussed in Chapter 6). In hatcheries using recirculated water or in smaller facilities using flow-through conditions, water treatment systems employing UV or ozone can often be practical. However, in pond culture, and in most large hatcheries operated under flow-through conditions, water treatment systems are usually impractical due to the high volumes of water involved. Fortunately, many of the disease problems resulting from fish–microorganism interactions can also be minimized by managing rearing conditions to meet the physiological needs of the fish and provide a low-stress environment. An example list of obligate and facultative fish pathogens and the disease problems they cause is given in Table 5.1.

Fish–Pathogen–Environment Relationship

Fish and their microbial pathogens interact with each other in a constantly changing relationship modulated by biological, chemical, and physical conditions in the aquatic environment to which both are connected by complex ties. The sum of these interactions is commonly termed the *fish—pathogen–environment relationship* and its outcome can be either continued fish health or infections followed by epizootic diseases and costly production losses. Fortunately, the fish culturist can often manage rearing conditions to favor the desired outcome—continued fish health. When the physical, chemical, and biological factors in the relationship

TABLE 5.1. Examples of facultative and obligate pathogens important to intensive fish culture.

Facultative Pathogens	Obligate Pathogens
Aeromonas hydrophila (motile *Aeromonas* septicemia)	*Aeromonas salmonicida* (furunculosis)
Vibrio anguillarum (vibriosis)	All fish viruses (IHN, CCV, VHS, IPN)
Flexibacter columnaris (columnaris disease)	*Flavobacterium branchiophilumb* (bacterial gill disease)
Edwardsiella ictaluri (enteric septicemia of catfish)	Protozoan parasites (*Ichthyobodo, Chilodonella, Ichthyophthirius, Myxobolus cerebralis*)
Fungi (*Branchiomyces* spp., *Saprolegnia* spp.)	*Renibacterium salmoninarum* (bacterial kidney disease)
Vibrio salmonicida (Hitra disease of farmed Atlantic salmon)	*Yersinia ruckeri* (enteric redmouth disease)

Source: Compiled from Wedemeyer and Goodyear (1984); Plumb (1994)

Notes: Flavobacterium branchiophilum originally thought to be a facultative pathogen, *Edwardsiella ictaluria* originally thought to be obligate.

are managed correctly, fish may be colonized by microorganisms but not harmed. If the factors involved in the relationship are not managed correctly, the outcome will be infections followed by epizootic diseases. Stressful fish cultural procedures, such as handling, can alter the outcome in favor of the pathogen by temporarily suppressing the fish's immune protection system. Water quality factors such as temperature and pH alter the outcome by affecting both the biochemical systems of the pathogen and the physiological condition of the fish. For example, bacterial growth and invasiveness are usually increased by warmer water temperatures, whereas the physiological condition of fish may be compromised by both the warmer temperature and the reduced concentration of dissolved oxygen available for respiration. There are exceptions, however. Colder water temperatures can increase disease problems caused by psychrophilic bacteria. For example, low temperature peduncle disease of salmonids (*Cytophaga psychrophilus* syn. *Flexibacter psychrophila*) abates above 13°C.

Proactive management of rearing conditions to prevent disease by establishing and maintaining a fish—pathogen–environment relationship favorable to the fish has long been a recommended method of disease control—particularly for intensive culture systems (Wedemeyer et al. 1976). Today, the increasing cost and decreasing availability of effective and legal antibiotics and other drugs and the still limited number of effective vaccines have made disease prevention through rearing condition management even more important. The degree of success depends on how many aspects of the fish—pathogen–environment relationship can be placed under the partial or complete control of the fish culturist. In intensive rearing systems, the major opportunities for control are usually water quality conditions, the stress of fish cultural procedures, and pathogen numbers and virulence. Pathogen control (with ozone or UV water treatment systems) is impractical in many hatchery situations, and the fish will unavoidably be exposed to varying numbers of both obligate and facultative pathogens on a more or less continuous basis. Thus, optimizing water quality factors and reducing the stress of fish cultural procedures to prevent adverse effects on physiological condition will usually be the most practical steps available.

Managing water quality factors to meet the physiological needs of the fish (discussed in Chapter 3) is the foundation of a favorable fish—pathogen–environment relationship. For example, the environmental gill disease syndrome of hatchery-reared salmonids begins insidiously as a generalized gill irritation probably due to poor water quality conditions. Secondary infections by facultative bacterial and fungal pathogens then occur followed by dramatic increases in the mortality rate (Klontz 1993).

Managing fish cultural procedures to mitigate the stress of handling, crowding, disease treatments, and transportation (discussed in Chapter 4) should be the next step. For example, fingerling channel catfish that are stressed by confinement for only 30 minutes suffer a significantly higher prevalence of *Edwardsiella ictaluri* infections than unstressed fish (Wise et al. 1993). The stress of social conflicts (dominance hierarchies) increases susceptibility to motile *Aeromonas* infections in subordinated fish (Peters et al. 1988). The pathogen also spreads through the organs and tissues more quickly in subordinated fish. Experience with warmwater pond culture has shown that the incidence of facultative bacterial infections in general can be considerably reduced if the fish are not subjected to handling stress in the early spring when water temperatures are rising. In coldwater fish culture, the facultative pathogens *Aeromonas, Pseudomonas,* and *Saprolegnia* spp. are continuously present in most water supplies, but again usually do not cause disease problems unless the fish are stressed by handling, crowding, or warm water temperatures.

Together with improving water quality conditions and reducing the stress of fish cultural procedures, the third opportunity for intervention in the fish—pathogen–environment relationship is to limit exposure by reducing the pathogen load in the water supply. For diseases spread primarily by waterborne (horizontal) transmission, one method of doing this is to use a closed (pathogen-free) water supply. Unfortunately, closed water sources (springs, wells) are not often available and, in any event, facultative organisms will quickly recolonize the water after it is exposed to the air in raceways or headboxes. In hatcheries using surface water supplies, preventing the initial entrance of pathogens is usually not possible. Two alternatives are available: inactivating pathogens in the incoming water by use of a disinfection system such as UV or ozone, and preventing the introduction of infected fish or eggs into the hatchery. Methods to prevent the introduction of infected fish or eggs by certifying brood stock as pathogen free are available (Thosen 1994) but will not be discussed here. Methods for inactivating pathogens using water treatment systems are discussed in Chapter 6.

Disease prevention through managing the rearing environment to establish a fish—pathogen–environment relationship that favors fish health is a proven and valuable technique, but it is not realistic to expect absolute success. For example, if virulent pathogens are allowed into the water supply, even in small numbers, disease problems are likely to occur regardless of the physiological condition of the fish. Diseases transmitted vertically (adult to progeny) are also difficult to prevent by this approach. Other factors being equal though, experience has shown that these epizootiological principles are valuable tools in minimizing or preventing fish disease problems in intensive rearing systems.

Infection into Disease: Mechanisms

As mentioned, diseases are only one outcome of the continuously occurring interactions between fish and aquatic microorganisms. These interactions are played out in a dynamic environment to which both fish and microorganism are connected by complex ties (Kabata 1984). The interactions between fish and microbial pathogens that may be harmless under natural conditions often result in disease problems in hatchery fish because of the added stress from the physical, chemical, or biological challenges inherent in intensive culture systems. Such challenges elicit a catecholamine and corticosteroid hormone cascade that brings about a series of cardiovascular, respiratory, and other secondary physiological changes intended to help the fish avoid or escape from the challenge in question. In intensive culture systems however, opportunities for avoidance or escape are minimal and the physiological changes usually do little to improve survival. They may become harmful if the stress is prolonged. The harmful effects include varying degrees of growth suppression, reproductive dysfunction, and immunosuppression. The effects of stress on normal resistance to infections are particularly important because the immunosuppression can linger for some time after the acute physiological changes have returned to prestress levels (Maule et al. 1989).

The physiological mechanisms that allow infections to become established and to spread when fish are stressed are not completely understood, but cortisol-induced immunosuppression is apparently a key feature of the process. The immune response to invading pathogens can also be adversely affected by nutritional deficiencies, but stress factors such as biological, chemical, and physical challenges are probably more important to fish in intensive culture systems. Both the specific and nonspecific aspects of the immune protection system[15] can be sup-

15. Defined in Chapter 2.

pressed. The corticosteroid/heat stress method originally developed by Bullock and Stuckey (1975) to identify rainbow trout that are asymptomatic carriers of furunculosis is a practical application of this phenomenon. The suspect fish are injected with prednisolone or other synthetic corticosteroid and then stressed by a water temperature of 18°C. If the fish are indeed latent carriers of *Aeromonas salmonicida*, the subsequent immunosuppression allows the bacteria to begin to reproduce and the resulting septicemia will quickly kill the fish. The specific mechanisms affected that allow new infections to become established or latent infections to progress into diseases are not completely understood, but include:

- Decreased serum bactericidal activity (complement[16], lysozyme)
- Impaired phagocytosis (macrophages, lymphocytes)
- Decreased leukocyte count (lymphocytopenia)
- Decreased antibody production by lymphocytes

Serum microbicidal activity, due to lysozyme and the complement system, and macrophage phagocytosis are the principal defense mechanisms of the nonspecific immune protection system. Both of these nonspecific defenses are adversely affected when fish are stressed. For example, carp subjected to low dissolved oxygen, starvation, or high salinity water suffer a significant decrease in serum bactericidal activity. The complement system and lysozyme activity are both depressed (Hajji et al. 1990). Lysozyme activity in rainbow trout remains low for at least 24 hours after the fish have been transported for only 2 hours (Möck and Peters 1990).

Macrophage phagocytosis is impaired by a variety of stress factors in channel catfish and rainbow trout. These include handling, transportation, confinement and even noise (Ellsaesser and Clem 1986; Nanaware et al. 1994). In most cases, both the ability to phagocytize (initial engulfment) and the intracellular killing of the phagocytized bacteria are adversely affected (Figure 5.3). The weakened ability to phagocytize bacteria is apparently due to the direct action of increased blood cortisol concentrations on the fluidity of the macrophage cell membrane. The failure to kill bacteria after they have been phagocytized is apparently also a side effect of hypercortisolemia. Two mechanisms are probably involved: failure of intracellular enzymes to disintegrate the ingested microbe after it is fused with the cytoplasmic phagosome, and interference with the killing action of nonspecific cytotoxins produced by the macrophage. The cytotoxins include hydrogen peroxide (H_2O_2) and the superoxide anion (O_2^-). Both are produced from oxygen during the respiratory burst that is stimulated when microbes attach to the plasma membrane during the phagocytic process. The macrophage protects itself from

16. Collective name for a serum protein system that nonspecifically lyses gram-negative bacteria and inactivates enveloped viruses.

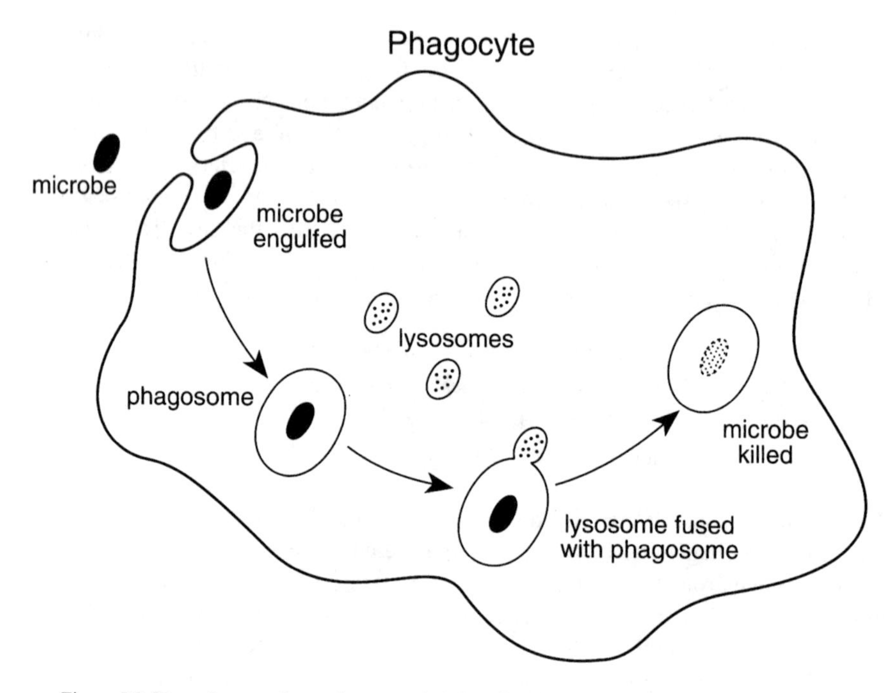

Figure 5.3. Normal macrophage phagocytosis is impaired by stress factors such as handling, transportation, confinement, and even noise. Both the ability to phagocytize (initial engulfment) and the intracellular killing of the phagocytized bacteria are adversely affected.

damage by destroying excess amounts of these highly reactive oxygen species with the enzymes catalase and superoxide dismutase[17] located in the cytoplasm. Phagocytized bacteria that are not killed can continue to metabolize and divide and will eventually kill the macrophage. In fish, effects on disease resistance seem to involve failure of the intracellular killing process more than failure of phagocytosis itself. For example, macrophages containing viable, apparently dividing bacteria can often be found in salmonids suffering from bacterial kidney disease (*Renibacterium salmoninarum*) or channel catfish with *Edwardsiella* septicemia (*Edwardsiella ictaluri*) infections (Blazer 1991).

Effects of stress on other important macrophage functions such as debris removal during tissue repair, uptake and processing of antigens, and the secretion of lysozyme and interferon may also be involved in allowing infections to become established and spread but are little understood at present. Evidence for the role of interferon in fish is indirect, but in rainbow trout, for example, recovery from VHS (viral hemorrhagic septicemia) infections often occurs spontaneously if the

17. Catalyzes the dismutation of the superoxide anion (O_2^-) to H_2O_2. Excess H_2O_2 is destroyed by the enzyme catalase.

water temperature is increased to 15°C or more. However, the virus develops quite normally in vitro at these temperatures. The thermodependence of the clinical disease may be partly due to the protective effects of faster interferon synthesis by leukocytes at warmer temperatures (de Kinklein et al. 1992).

Together with impaired phagocyte function, total numbers of circulating leukocytes are also reduced in stressed fish. The leukopenia (mainly lymphocytopenia) is also a side effect of hypercortisolemia and plays a major role in increasing susceptibility to microbial invasion and in allowing infections to spread (Pickering 1984; Barton and Iwama 1991).

In addition to decreasing the circulating leucocyte count, stress induced hypercortisolemia also interferes with antibody production by lymphocytes. For example, Atlantic salmon parr subjected to the stress of confinement for only 2 hours suffer impaired lymphocyte function and reduced production of specific antibody to *A. salmonicida* (furunculosis) (Thompson et al. 1993). The mechanism apparently involves suppression of cytokines such as interleukin-1 that are involved in antibody production (Kaattari and Tripp 1987). However, it has long been recognized that serum antibody titre does not necessarily correlate well with disease resistance in fishes—probably because of their greater reliance on the nonspecific mechanisms of the immune system for protection against invading microorganisms (Landolt 1989).

As mentioned, the mechanisms involved in the establishment and spread of infections also include a link to the nutritional status of the fish. As might be expected, the naturally occurring interactions between nutritional deficiencies, the immune system, and infectious agents are complex and studies have often yielded conflicting results. Nonetheless, a few general principles have been revealed.

Frank malnutrition usually facilitates infections because of general debilitation of physiological condition and the immune system. For example, starvation increases the prevalence of ceratomyxosis (*Ceratomyxa labracis* and *C. diplodae*) in cultured sea bass (Alvarez-Pellitero and Sitjà-Bobadilla 1993). One mechanism may be that starvation tends to reduce the nonspecific bactericidal activity of serum complement and lysozyme. Malnutrition can potentially result in antagonism to infection if the infectious agent is dependent on specific host enzyme systems or metabolites, or has a greater requirement for a particular dietary nutrient than the host.

Vitamin and micronutrient deficiencies are also usually synergistic to infections. Vitamin C, in particular, is widely considered to have generally beneficial effects on disease and stress resistance in both salmonids and channel catfish when fed at the basic requirement level of 50–100 ppm (mg/kg). Feeding megadoses of vitamin C (3,000 mg/kg) completely protected channel catfish against *Edwardsiella ictaluri* (enteric septicemia) infections (Li and Lovell 1985). Feeding vitamin C (and E) at higher than basic requirement levels has also been shown to have additional beneficial effects on the disease resistance of Atlantic and Pacific salmon

(Leith and Kaattari 1989; Hardie et al. 1990). Unfortunately, specific biochemical and physiological mechanisms by which vitamins and micronutrients act to affect disease resistance have been difficult to elucidate. For example, leukocyte bactericidal activity, migration, and phagocytic index (mean numbers of bacteria phagocytized per cell), are all unaffected in salmonids and channel catfish fed high levels of vitamin C (Thompson et al. 1993; Johnson and Ainsworth 1991).

The vitamin C status and nutritional state can also influence the spread of infections by affecting production and maintenance of repair tissue. Vitamin C and the sulfur-containing amino acids are required for the deposition of fibrin, collagen, and polysaccharides into the reticulum formed to isolate invading microbial pathogens by walling off infections. Deficiencies can inhibit the walling-off process. In addition, the walling-off process, especially the polysaccharide and protein (fibrin) components of the reticulum, is also susceptible to hydrolytic and proteolytic enzyme attack. A substantial increase in nonspecific plasma proteolytic activity can be stimulated by bacterial pathogens that produce endotoxins or by certain types of stressful challenges. Chronically stressful rearing conditions such as low DO, tend to decrease serum lysosome activity whereas acute challenges such as transportation or confinement increase it in both carp and Atlantic salmon (Hajji et al. 1990; Thompson et al. 1993). Thus, it is possible that acute stress may act synergistically with a vitamin C deficiency to facilitate the spread of invading pathogens through fish tissues.

B-complex vitamin and mineral deficiencies behave variably depending on the host and infectious agent. Pyridoxine (vitamin B_6 group) improves the resistance of salmonids to *Vibrio anguillarum* (vibriosis), *Aeromonas salmonicida* (furunculosis), and *Renibacterium salmoninarum* (bacterial kidney disease) infections (Hardy et al. 1979; Leith and Kaattari 1989). In the case of mineral deficiencies, low serum iron levels would be expected to reduce the invasiveness of bacterial pathogens such as *Vibrio anguillarum* (vibriosis), *Renibacterium salmoninarum* (bacterial kidney disease), and *Aeromonas salmonicida* (furunculosis) that use iron-chelating siderophores[18] to obtain the serum iron needed for their pathogenicity. Thus, farmed Atlantic salmon with high levels of serum iron are more susceptible to *Vibrio* infections than fish with lower iron levels (Ravndal et al. 1994). In contrast, dietary iron, zinc, or iodine deficiencies seem to increase the susceptibility of farmed Atlantic salmon to bacterial kidney disease (Plumb 1994). Micronutrients may also affect resistance to infections by affecting the process of phagocytosis. Adverse effects on phagocytosis could play a particularly critical role in fish because of their greater reliance on the nonspecific defenses of the immune protection system. For example, copper and zinc are required for optimum activity of superoxide dismutase, the enzyme that helps protect phagocytic

---

18. Organic chelating agents secreted into the environment that bind and transport iron back into the bacterial cell.

cells against damage by the cytotoxins they use to kill engulfed microorganisms. No work has been done with fish, but zinc deficiency in rats results in superoxide, free radical, and H_2O_2 accumulation in several phagocytic cell types (Kubow et al. 1986).

Dietary lipid composition may also affect macrophage function. Channel catfish fed diets formulated with fish oils rich in omega-3 unsaturated fatty acids produced macrophages that had much higher bactericidal activity than those from fish fed diets formulated with lipids high in saturated fat (Blazer 1991). One mechanism for such changes in phagocytic activity may be that the fatty acid composition of the macrophage cell membrane, which influences its fluidity, is easily altered by diet (Johnston 1988). The possibility of enhancing macrophage function by manipulating diet constituents is an attractive concept for managing the health of fish, although it is a remote possibility at present.

In summary, whether contact with a pathogen progresses into overt disease depends on the outcome of a complex series of interactions between the biochemical systems of the invading microorganisms and the physiological defense mechanisms of the fish. Both are modulated by biological, chemical, and physical conditions in the rearing environment. Bacterial and fungal pathogens, in particular, are often secondary invaders that cause diseases only if the stress of rearing conditions has adversely affected the physiological condition of the fish. Temperature changes, handling, low DO, transportation, crowding, and behavioral conflicts are commonly occurring examples of stressful rearing conditions. The physiological changes during the stress response collectively potentiate fish disease processes by interfering with the immune protection system. Aspects of both the specific (antibody) and nonspecific response are affected. Changes in phagocytosis and serum complement and proteolytic (lysozyme) activity associated with alterations in plasma steroid hormone ("stress hormone") concentrations are considered to be particularly significant effects because antibody titre does not relate well to protection for many fish diseases. A working knowledge of the mechanisms by which stressful physical, chemical, and biological challenges act to potentiate infectious processes is of considerable practical importance in disease prevention through rearing condition management. Postinfection events are particularly important in the prevention and control of fish disease problems due to facultative pathogens because it is usually impractical to remove these organisms from the water supply.

Stress-Mediated Diseases

Snieszko (1954) was among the first to point out that the incidence of certain freshwater fish diseases in Europe and North America tended to be seasonal, whereas in tropical climates no such relationship usually existed. That is, environmental stress (seasonal water temperature changes), in addition to pathogen exposure, was required for the initiation of epizootics. In succeeding years, it became clear that problems with bacterial diseases tended to increase in the spring because the growth and invasiveness of the pathogens involved responded more rapidly to the rising water temperatures than the immune defenses of the fish whose physiological condition had been debilitated by the overwintering environment (Wedemeyer et al. 1976). Since then, it has become well recognized that most of the infectious disease problems of hatchery (or wild) fish are multifactorial in nature and not simply the inevitable result of pathogen exposure as previously supposed (Wedemeyer and Goodyear 1984).

Because of the complexity of the interactions between the pathogen, the aquatic environment, and the physiological condition of the fish, not all the specifics of the physical, chemical, and biological stress factors associated with outbreaks have been defined. For fish in intensive culture however, the commonly occurring predisposing factors have been identified. These include stressful hatchery practices such as handling, size grading, crowding, and transportation in the presence of adverse water quality conditions such as low DO, high ammonia and carbon dioxide concentrations, and unfavorable temperatures. The chief pathogens involved in the stress-mediated diseases of hatchery fish include the external fungi, protozoan parasites, most of the facultative and obligate bacteria, and a few viruses. Detailed information on the role of environmental factors in the epizootiology of most microbial diseases can be found in Plumb (1994). A few of particular interest are discussed next.

Of all the stress-mediated fish disease problems identified to date, those due to facultative bacterial pathogens are probably the most frequently encountered. Diseases due to obligate pathogens can be stress-mediated as well, but are somewhat less frequent because these microorganisms are not always present in the rearing environment. The usual stress factors involved are fish cultural procedures and unfavorable water quality conditions. Infections such as the motile *Aeromonas* septicemias (*A. hydrophila* and others), bacterial gill disease (*Flavobacterium branchiophilum*), furunculosis (*A. salmonicida*), columnaris (*Flexibacter columnaris*), and vibriosis (*Vibrio anguillarum*) are classical examples of stress-mediated bacterial diseases (Wedemeyer and Goodyear 1984).

Epizootics of motile *Aeromonas* septicemia (MAS), a hemorrhagic septicemia caused by *A. hydrophila* and other motile species of the genus *Aeromonas*, have long been a ubiquitous problem in both cold- and warmwater fish culture. In warmwater pondfish culture, *A. hydrophila* is considered to be one of the primary bacterial pathogens. The major occurrence on catfish farms is during June, July, and August corresponding to the season of the year when pond conditions are least favorable for fish. Dissolved oxygen depletions are common during this period and water temperatures are rising. MAS infections in channel catfish usually do not occur unless physiological condition is debilitated by multiple stress factors (Plumb 1994). The fish cultural procedures identified as contributory include overcrowding, overfeeding, excessive pond fertilization, and algal die-offs after blooms (Walters and Plumb 1994). The usual water quality factors involved are low DO (1–2 mg/L), elevated carbon dioxide (5–10 mg/L), and elevated ammonia (1–2 mg/L). Disease problems are often delayed for 1–2 weeks after the fish are stressed by a combination of adverse water quality conditions and stressful fish cultural procedures (Plumb 1980).

In coldwater raceway culture, the most common stress factors are handling and crowding. However, in rainbow trout, an important coldwater species which is aggressive, the stress of behavioral conflicts alone can be sufficient to result in MAS in defeated individuals if the pathogen is present in the water (Peters et al. 1988).

Furunculosis, a bacterial septicemia of salmonids and a few cool- and warmwater fish, is another disease often correlated with stressful water quality conditions and fish cultural procedures. The etiologic agent, *A. salmonicida* is an obligate pathogen but disease outbreaks may not occur independently of stressful conditions. Kingsbury (1961) was the first to systematically correlate furunculosis outbreaks in hatchery rainbow trout with specific environmental conditions—water temperatures above 10°C, DO levels below 5.5–6.0 mg/L, handling for size grading and transportation, and crowding beyond species guidelines. In Pacific salmon, crowding in terms of the areal density (fish/m^2) as well as the related volume density (fish/m^3) has also been defined (Nomura et al. 1992). For example, the incidence of *A. salmonicida* infections in adult chum salmon held in shallow freshwater ponds at areal densities of about 15 fish/m^2 was about 12% whereas fish held at the reduced density of 5 fish/m^2 were disease free. Similarly, the prevalence of *A. salmonicida* in organs and tissues was significantly higher in adult chum salmon held at the marginal DO levels of 6–7 mg/L than in fish held in water with a DO of 10 mg/L.

Managing rearing conditions to maintain the water temperature below 15°C, the DO near saturation, fish loadings within species guidelines (given in Chapter 4), and minimizing handling during high-risk periods when water temperatures are rising are all helpful in preventing furunculosis outbreaks. Water temperatures of 15–20°C exacerbate furunculosis problems in salmonids. If the underlying environmental conditions are corrected, drug treatments will suppress outbreaks. However, surviving fish normally become *A. salmonicida* carriers. In the case of smolts carrying this pathogen, the stress of transfer into seawater will often activate the latent infection. The resulting furunculosis outbreak can greatly reduce early marine survival and cause serious economic losses. In the commercial aquaculture of Pacific and Atlantic salmon, the corticosteroid-heat stress test is sometimes used to identify smolts free of latent infections prior to transfer into seawater net pens for grow-out (Eaton 1988).

Bacterial gill disease (BGD) is a secondary infection by any of several filamentous bacteria that usually does not occur unless predisposing environmental conditions have first irritated the gill tissue. The principal infectious agent is

Flavobacterium branchiophilum (syn. *branchiophila*), but *Cytophaga* and *Flexibacter* spp. are also known to infect injured gill tissue. Death by asphyxiation occurs when masses of these filamentous bacteria growing on the gill surface fatally impede gas exchange. BGD is primarily a disease of intensively cultured freshwater salmonids but warmwater fish including carp, catfish, and eels can also be affected. The stress factors that predispose fish to BGD have not been completely defined but the loading rate, ammonia and DO concentrations, and suspended solids levels have all been implicated (Bullock et al. 1994). It has long been accepted that the main predisposing factor to bacterial invasion is subtle gill injuries caused by elevated ammonia levels and suspended feed particles in the 5–10 μm size range (Chapman et al. 1987). In intensive culture systems, most outbreaks of BGD are associated with management errors such as overfeeding, overcrowding, inadequate water flow rates, low DO, increased un-ionized ammonia levels, and the accumulation of suspended particulate matter that cause gill irritation. However, the factors in the host-pathogen-environment relationship that predispose fish to BGD seem to be more complex than for other stress-mediated diseases and outbreaks are not predictable. BGD occasionally occurs in fish held under rearing conditions that appear to be optimum. As a consequence, BGD has been difficult to reproduce under controlled laboratory conditions even when the experimental fish are severely stressed (Ferguson et al. 1991)

If the underlying environmental problems can be corrected, infected fish recover rapidly when the bacteria are removed from their gills with treatment chemicals such as chloramine-T[19] (salmonids: 10–15 mg/L, 1 hour). A necessity for repeated treatments is evidence that the underlying environmental stress factors have not been correctly identified.

Other common bacterial fish diseases for which there is a high risk of outbreaks when unfavorable water temperatures and/or other stressful environmental conditions occur include enteric redmouth disease (*Yersinia ruckeri*) in rainbow trout and enteric septicemia (*Edwardsiella ictaluri*) of channel catfish. The risk of enteric redmouth outbreaks is highest during the spring and summer when water temperatures are between 11–18°C (Romalde et al. 1994). Outbreaks of enteric septicemia of catfish (ESC), the leading bacterial disease of farmed channel catfish in the United States, are also related to environmental conditions. ESC generally occurs seasonally (late spring, fall) when water temperatures are in the 18–28°C range and changing rapidly. This seasonality suggests a cause and effect relationship, but outbreaks can also occur when environmental conditions appear to be highly favorable to the fish (Klesius 1992). Stressful fish cultural procedures such as handling, transportation, and confinement are also known to increase mortality from ESC (Wise et al. 1993). The cost of vaccines, and the recent de-

19. *N*-sodium-*n*-chloro-para-toluenesulfonamide, an antibacterial agent also used to disinfect drinking water and to sanitize equipment in the dairy and food industries.

velopment of *E. ictaluri* strains resistant to the only drug presently labeled for ESC control (Romet®) are persuasive reasons for developing a better understanding of the ESC–fish–aquatic environment relationship.

Parasite infestations are also strongly affected by environmental conditions. Low numbers of parasites living in a commensal[20] relationship on the body surfaces of fish are common occurrences. However, altered environmental conditions can allow parasite numbers to increase to the detriment of the fish. In warmwater pond culture, the respiration of uncontrolled growths of aquatic macrophyte plants can cause nighttime oxygen depletions and high dissolved CO_2 concentrations that are particularly likely to alter the harmless commensal relationships of the parasites *Argulus*, *Gyrodactylus*, and *Trichodina* spp. *Ichthyophthirius*, *Gyrodactylus*, and *Lernaea* spp. provide particularly graphic examples of commensal relationships that are altered by water temperature. Lack of water movement and overcrowding also contribute to mortality. Biologists must be able to recognize the ecological conditions that allow infestations by the common fish parasites to become harmful and be prepared to take preventive action through prophylactic formalin treatments or to treat as soon as infections develop.

The incidence of *Ichthyophthirius multifiliis* (Ich) in warmwater pond fish culture shows a distinct peak during the spring months, with a marked prevalence during April in North America. This parasite is probably present in low numbers on the fish all during the winter but the cold water temperatures retard development of an epizootic until spring when warming trends begin. Water temperatures typically begin to climb in April and can approach the optimal range of 21–24°C that favors the development of this parasite. In channel catfish culture, handling stress during April is also known to be a contributing factor. Ich is less common during the summer months, especially when water temperatures are above 27°C.

Gyrodactylus spp. epizootics on channel catfish farms in North America also show a peak occurrence during April although outbreaks are not unusual from January through July. Water temperature probably plays its most important role in determining the generation time of *Gyrodactylus* spp. At 20–24°C, only few days are required between the time eggs are produced and hatched, and the larvae locate a host fish and develop into adults. At water temperatures below about 10°C, this time may be extended to several months (Post 1987).

The season for *Lernaea* parasitism begins when water temperatures rise above 18°C. The optimum temperature range is 22–30°C. Warm weather in the early spring or late fall may extend the period of infection beyond the normal April to October season.

Noninfectious (physiological) fish disease problems can also be associated with improper management of rearing conditions. For example, environmental

20. A host–parasite relationship in which the parasite obtains food or other benefit from the host without either harming or benefiting it.

gill disease (gill necrosis), coagulated yolk disease of salmonid eggs and fry, swim bladder stress syndrome, and the life-threatening diuresis responsible for delayed mortalities after freshwater fish are transported are common fish health problems that can result from stressful rearing conditions. Most of these were discussed in Chapter 3. The so-called swim bladder stress syndrome can be a serious physiological problem in the intensive rearing of tilapia, striped bass, and marine perciform fishes such as the sea bream and sea bass. In physostomus and some physoclistous[21] species, young fish initially inflate the gas bladder by swallowing air at the surface. Larval fish may be prevented from doing this by the oil film left on the surface by some formulated diets. Hypoxic conditions may prevent initial inflation in physoclistic fish such as tilapia that do not initially swallow air at the surface (Summerfelt 1991). The survival rate of larval fishes with underinflated gas bladders is very low if they are challenged by any kind of stress such as handling, or low oxygen (Soares et al. 1994).

A summary of the common infectious and noninfectious fish disease problems currently thought to be associated with adverse rearing conditions is presented in Table 5.2.

Diseases as Indicators of Environmental Quality

Experience gained both from hatchery operations and from pollution monitoring programs in natural waters has shown that the occurrence of stress-mediated fish diseases can serve as useful early warning indicators that unfavorable conditions are developing in the aquatic environment (Wedemeyer and Goodyear 1984). These diseases can provide particularly useful information on the indirect effects of supposedly sublethal water quality conditions. For example, exposing rainbow trout to otherwise sublethal amounts of dissolved copper both reduces their normal resistance to infectious pancreatic necrosis virus infections and facilitates the horizontal transmission of enteric redmouth disease (*Y. ruckeri)*. Both these effects have the potential for causing a substantial indirect mortality (Hetrick et al. 1979; Hunter et al. 1980). Information on stress-mediated disease problems can also be helpful in evaluating the true biological costs and benefits associated with proposed water development projects. For example, fish ladders that provide access to previously blocked spawning habitat can inadvertently increase prespawning mortality from columnaris (*Flexibacter* sp.) infections in returning adult salmon if periods of high river water temperatures coincide with high fish densities in the ladders (Wedemeyer and Goodyear 1984).

Fish with noninfectious health problems, such as liver neoplasms and chronic inflammatory bile duct disease, can also provide information useful in interpreting environmental conditions. For example, fish with chronic inflammatory bile

---

21. No pneumatic duct between the esophagus and the gas bladder; discussed in Chapter 2.

TABLE 5.2. Environmental factors commonly associated with the occurrence of infectious and noninfectious fish diseases.

Fish Disease Problem	Predisposing Environmental Factors
Bacterial gill disease (*Flavobacterium* sp.)	Crowding; chronic low oxygen (4 mg/L for salmonids); elevated ammonia (more than 0.02 mg/L for salmonids); suspended particulate matter
Blue sac, hydrocele	Temperature; ammonia; crowding
Columnaris (*Flexibacter columnaris*)	Crowding or handling during warmwater periods if carrier fish are present
Environmental gill disease	Adverse rearing conditions, but contributory factors currently not well defined
Epithelial tumors, ulceration	Chronic, sublethal contaminant exposure
Fin erosion	Crowding; low DO; nutritional imbalances; chronic exposure to trace contaminants; high total suspended solids; secondary bacterial invasion
Furunculosis (*Aeromonas salmonicida*)	Low oxygen (<5 mg/L for salmonids); crowding; temperature; handling when pathogen carriers are present
Hemorrhagic septicemias, red-sore disease (*Aeromonas*, *Pseudomonas*)	External parasite infestations; ponds not cleaned; crowding; elevated ammonia; low oxygen; stress due to elevated water temperatures; handling after overwintering at low temperatures
Kidney disease (*Renibacterium salmoninarum*)	Water hardness less than about 100 mg/L (as $CaCO_3$); diet composition; crowding; temperature
Nephrolithiasis	Water high in phosphates and carbon dioxide
Parasite infestations	Overcrowded fry and fingerlings; low oxygen; excessive size variation among fish in ponds
Skeletal anomalies	Chronic, sublethal contaminant exposure; adverse environmental quality; PCB, heavy metals, Kepone, Toxaphene exposure; dietary vitamin C deficiency
Spring viremia of carp	Handling after overwintering at low temperatures
Strawberry disease (rainbow trout)	Uneaten feed; fecal matter with resultant increased saprophytic bacteria; allergic response
Sunburn	Inadequately shaded raceways; dietary vitamin imbalance may be contributory
Swim bladder stress syndrome	Oil films; hypoxia; salinity; other water quality
Vibriosis (*Vibrio anguillarum*)	Handling; DO <6 mg/L, especially at water temperatures of 10–15°C; salinity 10–15 ‰
White-spot, coagulated-yolk disease	Environmental stress: supersaturation >102–103%, temperature, metabolic wastes, chronic trace contaminant exposure

Source: Compiled from Wedemeyer and Goodyear (1984); Johnson and Katavic (1984); Sindermann (1988); Klontz (1993).
Note: Incidence of BKD may be inversely proportional to ionic composition, but not necessarily to total hardness (Fryer and Lannan 1993).

duct disease have been used as indicators of polluted areas in Lake Ontario, Canada. This disease results in badly obstructed bile ducts[22] and is often associated with secondary parasite infestations as well (Kirby and Hayes 1991).

In marine fishes, infectious diseases indicating that environmental quality has deteriorated beyond physiological tolerance limits include vibriosis (*V. anguillarum*), fin rot (syn. fin erosion; *Aeromonas* and *Pseudomonas* spp.), ulcer disease, viral lymphocystis, and ichthyophoniasis or other fungus infestations. *Ichthyophonus hoferi* is widely considered to be the most ubiquitous fungal pathogen of marine fish and epizootics of ichthyophoniasis have occurred repeatedly in herring populations of the western North Atlantic ocean—to the extent that population declines have occurred (Sindermann and Chenoweth 1993). Noninfectious diseases associated with otherwise sublethal water quality alterations in the marine environment include chromosomal and morphological abnormalities of eggs and larvae, skeletal abnormalities, skin ulcers, fin rot, epithelial papillomas, and neoplastic and preneoplastic liver nodules (Sindermann 1988). Thus, statements about the habitat requirements of fishes in fresh or marine waters must include information on the fish—pathogen–environment relationship to be complete.

It is possible for both acute and chronic stress-mediated diseases to result in population declines if poor health reduces the survival, longevity, growth, or reproductive success of sufficient numbers of individual fish. Such effects are most likely to act through the density-dependent compensatory processes that passively regulate population size through negative feedback (Wedemeyer and Goodyear 1984).

The size of a population at any particular time is determined by the balance between birth and death rates. If losses due to mortality are offset by reproductive success, the population will be stable. If losses from mortality are less than the rate of recruitment of younger fish to succeeding life stages, the population will increase. If the mortality rate exceeds the recruitment rate, the population will decline. A growing population will eventually experience resource limitations[23] and mortality rates will then increase and/or reproductive success will decrease until a population size is reached that environmental conditions can support. If a fish population starts to decline, reproductive success (fecundity or larval survival) will usually begin to increase. Mortality of individual fish due to disease instead of senescence (or fishing) only alters the timing of their inevitable death. Thus, an increase in the death rate from stress and disease that reduces the number of adult fish surviving to spawn will normally also be compensated for by an increase in reproductive success. The ability of populations to compensate by these so-called density-dependent processes is limited, and when part of the compensatory reserve has been so used, the tolerance of the affected population to subsequent

22. Cholangiohepatitis and cholangiofibrosis.

23. The resource in shortest supply, usually food, physical habitat, or water quality.

stress factors (such as increased fishing pressure) will be correspondingly reduced.

A number of density-dependent mechanisms may be involved in the compensatory response to the direct or indirect effects of stress-mediated diseases and reliable predictions of likely population-level damage can be difficult to obtain. Merely documenting a particular incidence of mortality does not necessarily establish that the fish population as a whole will be permanently affected, even if the mortality was substantial. Even in cases where population declines are in fact occurring, they may be time-consuming to document because normal year-to-year variations in population abundance are sometimes quite large. Assuming that the population has actually been affected, whether it will recover or not depends on the nature and intensity of the mechanisms that control the relation between stock size and recruitment. In situations where stress and disease have affected physiology or behavior to the extent that normal growth or mortality patterns are directly altered, the ultimate response of the population will still be influenced by the outcome of all the other intra- and interspecific interactions that indirectly influence the rates of survival, growth, and reproduction of the affected individuals. It is entirely possible for a population to change very little, or to increase, if the density-dependent processes are sufficiently intense.

Even in the simplest case where diseases result in direct mortalities and no higher level of biological organization is directly affected, the requirements for establishing a cause and effect relationship for the role of stress and disease in population declines can be demanding. Several kinds of biological information are required. First, the nature, timing, and intensity of the density-dependent processes that regulate the size of the population in question must be understood. Second, the timing and magnitude of any increased mortality attributable to disease must be determined. Information on timing is important because the outcome of increased mortality due to diseases will usually be more severe if it occurs early rather than late in the life cycle (Shuter 1990). Information on magnitude must include both the direct and indirect mortality.

Direct mortality resulting from disease-induced changes in physiological function is usually easy to recognize and generally amenable to field measurement. However, stress and disease can also have chronic physiological consequences that result in changes in the indirect mortality rate through effects on other species in the community. For example, the interactions of diseased fishes with other trophic levels or with the aquatic environment may be modified. The indirect mortality resulting from such interactions may be all but impossible to quantify by presently available field-sampling techniques.

One such indirect effect that can at least be estimated is effects on predator avoidance. If poor health results in increased predation on the affected fish, the indirect mortality in the population may rise substantially above its normal level. Thus, chronic diseases such as the liver neoplasms found in groundfish that are exposed to contaminated bottom sediments may reduce population numbers to a greater extent than might be predicted on the basis of their direct effects on individual mortality. Again, the ultimate significance of indirectly caused mortality will also depend on its timing and magnitude with respect to the density-dependent processes that regulate population size. The death of adult fish toward the end of their normal life expectancy may be less important to the population than mortality at early life stages.

Another contributor to indirect mortality is growth reduction. Slower growth may be due to several factors including changes in feeding behavior or food conversion efficiency. A reduced growth rate during the early life stages is usually more important as an indirect source of mortality than slower growth in adult fish. Seemingly minor reductions in larval growth rates can profoundly increase total mortality because survival at early life stages is usually strongly size dependent. As an example, if the normal daily mortality rate for a larval fish is 0.25 during the 20 days from hatching to a particular size and slower growth increases this time to only 23 days, the surviving population size will be about 34% smaller. If stress and disease act to decrease the normal daily survival rate by other effects as well (in addition to reduced growth), then the total increase in mortality can be much greater.

Data on the amount that stress and disease reduces reproductive success is also needed. Poor fish health can directly affect reproduction through several mechanisms including effects on oogenesis and spermatogenesis, ovulation and spermiation, spawning, fertilization, and (in viviparous fish) embryonic development (Donaldson 1990). Poor health can also have indirect effects on reproductive success. For example reduced growth can both diminish the number of eggs produced by spawning females and delay the age at which sexual maturity is attained. Information on both direct and indirect effects on reproduction is needed because the total lifetime egg production by females will be reduced by both. A compensatory response such as increased survival of the fewer eggs produced would be required to allow such populations to remain stable.

Finally, information must be provided on the extent to which stress and disease reduce normal physiological capacity to cope with fluctuations in environmental conditions such as temperature, oxygen, and salinity. If physiological tolerance is affected, spatial and temporal variations in temperature, oxygen, or salinity will likely result in spatial and temporal variations in the survival of affected fish. For example, anadromous salmonid smolts with viral erythrocytic necrosis (VEN) infections have less tolerance to low DO conditions than uninfected fish (MacMillan et al. 1980).

In summary, establishing the population-level significance of stress-mediated diseases requires quantitative information on changes in the ability of affected fish to survive, grow, and reproduce in their local environment together with knowledge of the density-dependent processes that regulate population size. The problem is complicated by the fact that environmental alterations resulting in poor health in one species will often result in poor health in other members of the community as well. Thus, quantitative information on interactive effects may also be required. Establishing an unequivocal cause-and-effect relationship would require information on the dynamics of the entire ecosystem of which the affected population is a part. The difficulties in designing and conducting the research needed to evaluate the effects of fish diseases on population stability can be substantial unless the mortality is obviously catastrophic. In general, controlled studies with appropriate experimental manipulations are impractical or even impossible to carry out. For example, epizootics of the ubiquitous fungal pathogen *Ichthyophonus hoferi* are widely considered to have reduced the abundance of Gulf of Saint Lawrence herring populations, but the environmental factors contributing to the initiation of disease outbreaks have yet to be fully defined (Sindermann and Chenoweth 1993). Research that takes advantage of planned or ongoing land and water use projects, such as cleanup of contaminated dredge spoil areas or improved treatment of effluents discharged into natural waters containing affected fish populations, is one way to make incremental progress.

MANAGING BIOLOGICAL INTERACTIONS TO PREVENT DISEASES

Because fish diseases usually require a predisposing stressful condition in the aquatic environment, together with the concurrent presence of a pathogenic agent and a susceptible host, experienced fish culturists take precautions to prevent the simultaneous occurrence of all three factors. These precautions, in a general sense, constitute the disease management program of a hatchery, fish farm, or other aquaculture facility. The specific actions involved in disease management include (1) problem analysis, (2) minimizing pathogen exposure, (3) environmental modification, (4) immunization, and (5) prophylactic drug treatments (Ahne et al. 1989). The first two of these actions are discussed briefly here and in more detail in Chapter 6.

Problem Analysis

An important first step in the disease management process is recognizing whether health and disease problems have resulted from an adverse fish–pathogen–environment relationship or from other factors such as toxicants or an oxygen depletion. It is vital that fish producers visually check their ponds or raceways daily for signs of unfavorable environmental conditions as indicated by fish piping (gasping) at the surface in the early morning, sudden changes in the water color, musty odors, heavy blooms of algae, and slicks or streaks at the surface. If losses are occurring, valuable information can sometimes be obtained by examining the mortality pattern. Losses that occur at a low rate over several weeks may indicate poor environmental conditions, external parasite infestations, or possibly a low virulence bacterial infection. Mortality that begins slowly and then rises sharply within 5–7 days is often indicative of highly virulent bacterial or viral pathogens. An abrupt mortality (within 24 hours) is indicative of an oxygen depletion or other event that made the rearing environment acutely toxic such as a pesticide spill. These mortality patterns are summarized diagrammatically in Figure 5.4.

Bacterial epizootics usually follow a common sequence of events. Most often only the fish present in a pond or raceway will be killed. In some cases, however, tadpoles and frogs will also be infected. If the epizootic progresses slowly, external lesions characteristic of the infection will usually be visible. In epizootics caused by highly virulent pathogens, the fish may die very quickly and external lesions may be absent. Such cases are not common but do occur. Acute mortalities due to bacterial epizootics can be differentiated from fish kills resulting from oxygen depletions by the fact that the latter generally begin during the early morning hours while it is still dark. Large fish will die first because of their higher oxygen requirements (see Chapter 2). Aquatic vegetation may also die and conspicuous changes in the odor and color of the water will frequently also be seen. Zooplankton will usually be survive. There is often a species selectivity in plankton mortal-

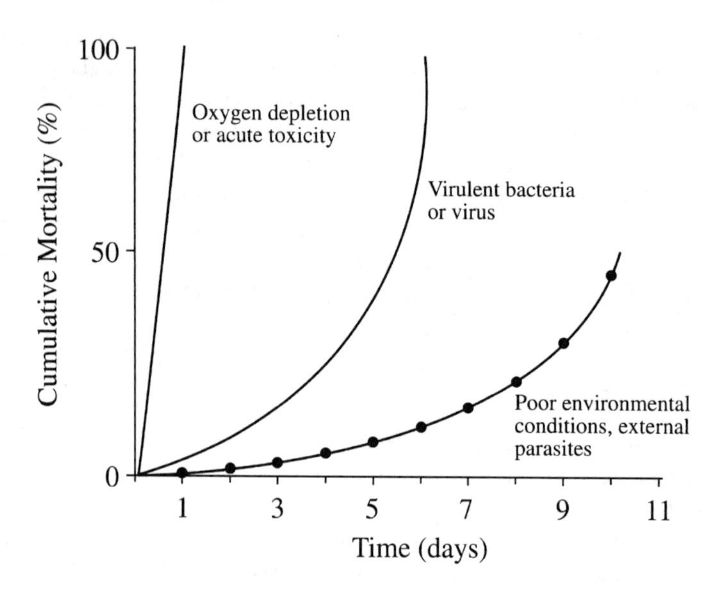

Figure 5.4. Mortality patterns associated with three major categories of fish health problems in intensive aquaculture.

ity with those species having the highest oxygen requirements dying first. The die-off usually ends shortly after sunrise when photosynthesis by aquatic plants begins to supply oxygen.

The fact that abrupt fish kills may also be due to toxicants should not be overlooked. Mortalities may begin at any hour and will continue through daylight and darkness. In pond culture, zooplankton, snakes, turtles, clams, and frogs may also be affected. Small fish will usually die first, usually exhibiting clinical signs such as convulsions, loss of equilibrium, and toxicosis. Some insecticides cause the fish to exhibit behavioral signs such as a forward thrusting of the pectoral fins; others may cause curvature of the spine (lordosis or scoliosis). A lethal toxicant such as cyanide may cause no observable outward signs but an examination of the internal organs will reveal massive internal hemorrhaging. Differential species sensitivity will be evident during the early stages of the kill but, in the case of highly toxic substances, mortalities will quickly progress through the entire fish community. Water conditions may appear normal with no particularly noticeable change in color or odor (Hunn and Schnick 1990).

Fish killed by an oxygen depletion may appear much the same as fish killed by toxicants except for flared opercles. As a consequence, no evidence should be overlooked. In intensive culture facilities, oxygen depletions are most often caused by mechanical problems that cause water failure or by human error. In pond culture, the development of heavy algal blooms due to high aquatic fertility levels brought about by dissolved minerals are a frequent cause. Algal blooms, coupled with the subsequent decomposition of the organic matter, increase the demand for dissolved oxygen above the replacement rate and lower the dissolved oxygen content of the water. If the rate of oxygen consumption exceeds the rate of replenishment by photosynthesis, or the oxygen budget for the water mass, an oxygen depletion results.

If the conditions that bring about oxygen depletions are understood, corrective actions by the fish culturist are more likely to be successful. Dissolved oxygen in the bottom pond water is the first to be depleted, so measuring oxygen 2 feet below the surface and near the bottom of the pond, just before sunrise, is one of the best methods for determining the overall dissolved oxygen status. When the DO 2 feet below the surface of the water declines to 2 ppm or less (warmwater pond fish) or 5 ppm in the case of salmonids, or fish are surfacing and showing signs of distress, emergency corrective action should be taken. Immediate flushing with clean well-aerated water is the best solution. If this is not possible, the oxygen-depleted water should be drained from the pond bottom during the day and replaced with freshwater before nightfall. If the freshwater is from a well, it may be of low oxygen content and also supersaturated with nitrogen. Spray aeration using commercially available surface aerators or paddle wheel units will help correct both problems. If water is drawn from an adjacent pond, it should be pumped from the surface—never from near the bottom—and sprayed into the new pond, or the flow directed over perforated metal plates or baffles. Water replacement or exchange should be continued until danger of an oxygen deficit has passed. If these measures are not possible and the fish appear to be in danger of immediate asphyxiation, applications of potassium permanganate ($KMnO_4$) to achieve a concentration of 2–6 ppm may provide temporary relief. Potassium permanganate does not directly supply DO but oxidizes organic matter and any sulfides and nitrites present that otherwise would consume dissolved oxygen needed by the fish (Huner and Dupree 1984). A continuous flow of some oxygenated freshwater to both increase the DO and reduce pond fertility by dilution is another solution. Fish should not be fed until the low DO conditions have been corrected because of the increased oxygen consumption this causes (discussed in Chapter 2).

If the dissolved oxygen concentration is less than 3 ppm at daybreak and does not rise during the day (bright sunlight), it can be assumed that either there is no photosynthetic activity (i.e., no living algae remain) or that bacterial decay is using the oxygen faster than it can be produced by the algae. In this situation, or when the oxygen content is known to be low (but the fish are not in obvious danger), applying 25 to 50 lb of calcium hydroxide[24] ($Ca(OH)_2$) per surface acre will help lower the high dissolved CO_2 that is usually present under these conditions. This enables the fish to use the remaining DO more efficiently by improving blood oxygen transport via the Bohr effect discussed in Chapter 2. A concentration of 21 ppm of calcium hydroxide will remove about 25 ppm CO_2. Treatments may be repeated as needed to reduce the dissolved CO_2 to below about 10 mg/L. The minimum amount of lime should always be used because it will usually cause the water pH to rise thus exacerbating ammonia toxicity problems (Tucker and Robinson 1990).

Too few algae and zooplankton to provide fish food organisms, or to shade out unwanted rooted aquatic plants, can also be a problem. In this case, fertility should be increased by adding 20 lb of ammonium phosphate or other fertilizer per surface acre. Any formulation rich in nitrogen and phosphorus (e.g., 10–20–0 or 16–20–0 N-P-K) will produce the fertility required to rapidly produce a bloom. However, do not add fertilizer if an algal bloom is already underway. Excessive fertilization will result in high levels of organic matter that, in turn, can lead to problems with fungal diseases. Outbreaks of gill branchiomycosis, caused by fungi of the genus *Branchiomyces* spp., can be a particular problem in this regard (Wedemeyer et al. 1976).

An epizootic due to a parasite infestation can be expected to have an extended course with fish dying over a prolonged period. The loss typically affects a limited number of species. The fish usually appear weak or listless and often will be emaciated as well. Generally, the die-off will affect a few fish at the outset and the

24. Sold commercially as slaked lime or hydrated lime.

mortality rate will steadily increase with time. Small fish usually die first and aquatic animals such as toads or frogs are not affected. Infestations by parasites that require an intermediate host develop more slowly but are rare in intensive culture systems because the concrete raceways or ponds tend to exclude intermediate hosts. As a consequence, parasite disease problems in intensive fish culture facilities also develop quickly because normally only parasites with direct life cycles are involved. The key to control lies in recognition of conditions favoring development of the parasite, early detection, and prompt treatment (normally with formalin).

Pathogen Exposure

Disinfecting raceways at least once a year by drying is an inexpensive method of reducing the buildup of fish pathogens residing in organic matter and on concrete surfaces. In the case of dirt ponds, thorough drying followed by disking and leveling of the pond bottom facilitates air oxidation and mineralization of organic substances that would otherwise contribute to septic conditions or heavy algal blooms when the ponds are refilled. Drying dirt ponds also helps to control parasitic diseases. Protozoans, monogenetic trematodes, leeches, and copepods all can be eradicated by this practice. Areas that do not dry completely should be treated with calcium hydroxide (slaked lime) or calcium oxide (quicklime) at about 5 lbs/100ft^2. Calcium hypochlorite (HTH) can also be applied to areas of wet soil but guidelines for the application rate are not available. The usual recommendation for treating pond water is sufficient HTH to achieve a concentration of about 10 ppm chlorine with a contact time of 2–3 days (see discussion in Chapter 3).

Wild fishes or their eggs brought onto the premises of hatcheries or fish farms can be another preventable source of pathogen exposure. The diseases carried by wild fish are often in the form of latent, and thus asymptomatic, bacterial and viral infections. The low population density of fishes in most natural waters serves as a biological restraint on the horizontal transmission of such pathogens. Stresses due to capture and handling can allow unsuspected subacute infections to progress to the clinical stage and then to an epizootic within days after wild fish are brought into a hatchery. If fish from the wild must be used, they should be quarantined and steps taken to inactivate bacterial pathogens and external parasites by formalin and antibiotic treatments. Feeding commercially available Terramycin® (oxytetracycline) medicated diets for two weeks or injecting individual fish with Terramycin at a doseage of 25 mg per pound of body weight often proves effective in controlling motile *Aeromonas* septicemias and other bacterial infections. Formalin treatments (200 ppm, 1 hour) to control external parasites should also be mandatory.

Another technique that will reduce the probability of introducing parasites and other pathogens when wild broodfish must be used in hatcheries is to treat for external parasites with formalin and then use artificial spawning techniques, keeping the fertilized eggs separated from the adults. For example, removing extensively cultured carp from spawning ponds before the eggs hatch is successful in controlling *Ichthyophthirius* infestations. Similar results have been obtained in catfish culture. The importance of treating captive broodstock to prevent the exposure of juvenile fish to parasites also cannot be over emphasized. Egg disinfection with iodophores such as Betadine will further reduce the incidence of disease transmission and should be considered to be a routine hatchery management practice. Failure to use these techniques has historically been a common cause of *Ichthyophthiriasis* epizootics on carp farms worldwide (Wedemeyer et al. 1976).

Reduced immunity to pathogens can be an important reason for the poor survival that sometimes occurs when hatchery fish are stocked into natural waters. Feeding Terramycin medicated feed at the rate of 2.5–3.5 grams TM-50 per 100 lb of fish per day for 10 days prior to release will temporarily protect most cold- and warmwater fish from activation of their own latent infections and from bacterial pathogens in their new environment. For example, feeding TM-50 medicated diets to hatchery smolts for two weeks prior to release helps prevent vibriosis mortality in anadromous salmonids which rear for a time in estuaries during their seaward migration

In summary, biologists must consider all aspects of the fish—pathogen–environment relationship to acquire the information needed to maximize health and physiological condition. Seemingly minor factors of weather, water flow, nutritional status, and water chemistry may play important causative roles in the development of fish disease problems. The following unique situation originally cited in Wedemeyer et al. (1976) is presented as a summarizing example. On the third day of warm winter weather, white catfish held in a 0.1-acre pond began to feed heavily on chironomid larvae. The temperature then fell from a daytime average of about 30°C to below freezing, reaching about –3°C by the morning of the fourth day. The rapid temperature change was accompanied by strong winds that speeded cooling of the pond waters. During the morning of the fifth day, large numbers of moribund catfish fingerlings were seen floating at the surface, some in rigor. The fish had distended abdomens, were unable to swim beneath the surface, and soon began dying in large numbers. The distended abdomens suggested ascites due to a facultative bacterial infection but necropsy revealed that the stomach was filled with gas. The fish were in fact bloated. The drop in water temperature had apparently been so rapid that the fish were unable to finish digesting the food filling the stomach and gut. Bacterial fermentation then began and generated sufficient gas to force the fish to the surface. Transferring the surviving fish to warmer water slowed the mortality rate and many of the fish recovered. However, a significant mortality from both parasitic and bacterial infections later occurred because of the stress of the sudden temperature change and handling.

REFERENCES

Abbott, J. C., and L. M. Dill. 1989. The relative growth of dominant and subordinant juvenile steelhead trout (*Salmo gairdneri*) fed equal rations. Behavior 108:104–111.

Ahne, W., J. R. Winton, and T. Kimura. 1989. Prevention of infectious diseases in aquaculture. Journal of Veterinary Medicine 36:561–567.

Alvarez-Pellitero, P., and A. Sitjà-Bobadilla. 1993. *Ceratomyxa* spp. (Protozoa: Myxosporea) infections in wild and cultured sea bass, *Dicentrarchus labrax*, from the Spanish Mediterranean area. Journal of Fish Biology 42:889–901.

Austin, B., and D. Allen-Austin. 1985. Bacterial pathogens of fish. Journal of Applied Bacteriology 58:483–506.

Austin, B., and D. Allen-Austin. 1987. Bacterial Fish Pathogens: Disease in Farmed and Wild Fish. Ellis Horwood, London.

Barton, B. A., and G. K. Iwama. 1991. Physiological changes in fish from stress in aquaculture with emphasis on the response and effects of corticosteroids. Annual Review of Fish Diseases 1:3–26.

Blazer, V. S. 1991. Piscine macrophage function and nutritional influences: A review. Journal of Aquatic Animal Health 3:77–86.

Bullock, G. L., and H. M. Stuckey. 1975. *Aeromonas salmonicida* detection of asymtomatically infected trout. Progressive Fish-Culturist 37:237–245.

Bullock, G., R. Herman, J. Heinen, A. Noble, A. Weber, and J. Hankins. 1994. Observations on the occurrence of bacterial gill disease and amoeba gill infestation in rainbow trout cultured in a water recirculation system. Journal of Aquatic Animal Health 6:310–317.

Chapman, P. E., J. D. Popham, J. Griffin, and J. Michaelson 1987. Differentiation of physical from chemical toxicity in solid waste fish bioassay. Water, Air, and Soil Pollution 33:295–308.

Cooper, E. L., G. Peters, I. Ahmed, M. Faisal, and M. Choneum. 1989. Aggression in Tilapia affects immunocompetent leucocytes. Aggressive Behavior 15:13–22.

de Kinklein, P., M. Dorson, and T. Renault. 1992. Interferon and viral interference in viroses of salmonid fish. Pages 241–249 *in* Proceedings of the OJI International Symposium on Salmonid Diseases. Hokkaido University Press, Sapporo, Japan.

Donaldson, E. M. 1990. Reproductive indices as measures of the effects of environmental stressors in fish. American Fisheries Society Symposium 8:109–122.

Eaton, C. A. 1988. The use of stress testing to prevent the movement of salmonids that are latently infected with furunculosis. Bulletin of the Aquaculture Association of Canada 88:73–75.

Ejike, C., and C. B. Schreck. 1980. Stress and social hierarchy rank in coho salmon. Transactions of the American Fisheries Society 109:423–426.

Ellsaesser, C. F., and L. W. Clem. 1986. Haematological and immunological changes in channel catfish stressed by handling and transport. Journal of Fish Biology 28:511–521.

Ferguson, H. W., V. E. Ostland, P. Byrne, and J. S. Lumsden. 1991. Experimental production of bacterial gill disease in trout by horizontal transmission and by bath challenge. Journal of Aquatic Animal Health 3:118–123.

Fryer, J. L., and C. N. Lannan. 1993. The history and current status of *Renibacterium salmoninarum*, the causative agent of bacterial kidney disease in Pacific salmon. Fisheries Research 17:15–33.

Grant, J. W. A. 1993. Whether or not to defend? The influence of resource distribution. Pages 1–4 *in* F. A. Huntingford and P. Torricelli (eds.), Behavioral Ecology of Fishes 23:137–153.

Hajji, N., H. Sugita, S. Ishii, and Y. Deguchi. 1990. Serum bactericidal activity of carp (*Cyprinus carpio*) under supposed stressful rearing conditions. Bulletin of the College of Agriculture and Veterinary Medicine, Nihon University, Nichidai-Nojuho 47:50–54.

Hardie L. J., T. C. Fletcher, and C. J. Secombes. 1990. The effect of vitamin E in the immune response of the Atlantic salmon (*Salmo salar*). Aquaculture 87:1–13.

Hardy, R. W., J. E. Halver, and E. L. Brannon. 1979. The effect of dietary pyridoxine levels on growth and disease resistance of chinook salmon. Pages 253–260 *in* J. Halver and K. Tiews (eds.), Finfish Nutrition and Fish Feed Technology, vol. I. Heenemann Verlagsgesellschaft, Berlin.

Hetrick, F. M., M. D. Knittel, and J. L. Fryer. 1979. Increased susceptibility of rainbow trout to infectious hematopoietic necrosis virus after exposure to copper. Applied Environmental Microbiology 37:198–201.

Huner, J. V., and H. J. Dupree. 1984. Pond management. Pages 17–43 *in* H. Dupree and J. Huner (eds.), Third Report to the Fish Farmers: The Status of Warmwater Fish Farming and Progress in Fish Farming Research. U.S. Fish and Wildlife Service, Washington, D.C.

Hunn, J. R., and R. A. Schnick. 1990. Toxic Substances. Pages 19–30 *in* F. Meyer and L. Barclay (eds.), Field Manual for the Investigation of Fish Kills. Resource Publication 177, Fish and Wildlife Service, U.S. Department of the Interior, Washington, D.C.

Hunter, V. A., M. D. Knittel, and J. L. Fryer. 1980. Stress-induced transmission of *Yersinia ruckeri* infection from carriers to recipient steelhead trout *Salmo gairdneri* Richardson. Journal of Fish Diseases 3:467–472.

Johnson, D. W., and I. Katavic. 1984. Mortality, growth, and swim bladder stress syndrome of sea bass (*Dicentrarchus labrax*) larvae under varied environmental conditions. Aquaculture 38:67–78.

Johnson, M. R., and A. J. Ainsworth. 1991. An elevated level of ascorbic acid fails to influence the response of anterior kidney neutrophils to *Edwardsiella ictaluri* in channel catfish. Journal of Aquatic Animal Health 3:266–273.

Johnston, P. V. 1988. Lipid modulation of immune responses. Pages 37–86 *in* R. K. Chandra (ed.), Nutrition and Immunology. R. Liss, New York.

Kaattari, S. L., and R. A. Tripp. 1987. Cellular mechanisms of glucocorticoid immunosuppression in salmon. Journal of Fish Biology 31 (Supplement A):129–132.

Kabata, Z. 1984. Diseases caused by metazoans: crustaceans. Pages 321–399 *in* O. Kinne (ed.), Diseases of Marine Animals, vol. 4, part 1: Introduction, Pisces. Biologische Anstalt Helgoland, Hamburg, FRG.

Kingsbury, O. R. 1961. A possible control of furunculosis. Progressive Fish-Culturist 23:136–138.

Kirby, G. M., and M. A. Hayes. 1991. Significance of liver neoplasia in wild fish: Assessment of pathophysiological responses of a biomonitor species to multiple stress factors. Pages 106–116 *in* A. Niimi and M. Taylor (eds.), Proceedings of the Eighteenth Annual Aquatic Toxicology Workshop. Department of Fisheries and Oceans, Ottawa, Ontario, Canada.

Klesius, P. 1992. Immune system of channel catfish: an overture on immunity to *Edwardsiella ictaluri*. Annual Review of Fish Diseases 2:325–338.

Klontz, G. W. 1993. Environmental requirements and environmental diseases of salmonids. Pages 333–342 *in* M. Stoskopf (ed.), Fish Medicine. W. B. Saunders, Philadelphia, Pennsylvania.

Knights, B. 1987. Agonistic behavior and growth in the European eel, *Anguilla anguilla* L, in relation to warmwater aquaculture. Journal of Fish Biology 31:265–276.

Kubow, S., T. M. Bray, and W. J. Bettger. 1986. Effects of dietary zinc and copper on free radical production in rat lung and liver. Canadian Journal of Physiology and Pharmacology 64:1281.

Landolt, M. L. 1989. The relationship between diet and the immune response of fish. Aquaculture 79:193–206.

Leith, D., and S. Kaattari. 1989. Effects of vitamin nutrition on the immune response of hatchery-reared salmonids. Final Report U.S. Department of Energy, Bonneville Power Administration, Division of Fish and Wildlife, Portland, Oregon

Li, Y., and R. T. Lovell. 1985. Elevated levels of dietary ascorbic acid increase immune responses in channel catfish. American Journal of Nutrition 115:123–131.

MacMillan, J. R., D. Mulcahy, and M. Landolt. 1980. Viral erythrocytic necrosis: some physiological consequences of infection in chum salmon (*Oncorhynchus keta*). Canadian Journal of Fisheries and Aquatic Science 37:799–804.

Maule, A. G., R. A. Tripp, S. L. Kaattari, and C. D. Schreck. 1989. Stress alters immune function and disease resistance in chinook salmon (*Oncorhynchus tshawytscha*). Journal of Endocrinology 120:135–142.

Mesa, M. G. 1991. Variation in feeding, aggression, and position choice between hatchery and wild cutthroat trout in an artificial stream. Transactions of the American Fisheries Society 120:723–727.

Möck, A., and G. Peters. 1990. Lysozyme activity in rainbow trout, *Oncorhynchus mykiss* (Walbaum), stressed by handling, transport, and water pollution. Journal of Fish Biology 37:873–885.

Nanaware, Y. K., B. I. Baker, and M. G. Tomlinson. 1994. The effect of various stresses, corticosteroids, and antigenic agents on phagocytosis in the rainbow trout *Oncorhynchus mykiss*. Fish Physiology and Biochemistry 13:31–40.

Nomura, T., M. Yoshimizu, and T. Kimura. 1992. An epidemiological study of furunculosis in salmon propagation. Pages 187–193 *in* T. Kimura (ed.), Salmonid Diseases. Hokkaido University Press, Hokodate, Japan.

Norman, L. 1987. Stream aquarium observations of territorial behavior in young salmon (*Salmo salar*) of wild and hatchery origin. Report of the Salmon Research Institute (Laxforskningsinstitute), Sweden.

Peters, G., and L. Quang Hong. 1984. The effect of social stress on gill structure and plasma electrolyte levels of European eels (*Anguilla* L.). Verfahren Deutsch Zoologisch Gesellschaft 7:318–322.

Peters, G., M. Faisal, T. Lang, and I. Ahmed. 1988. Stress caused by social interaction and its effect on susceptibility to *Aeromonas hydrophila* infection in rainbow trout *Salmo gairdneri*. Diseases of Aquatic Organisms 4:83–89.

Pickering, A. D. 1984. Cortisol-induced lymphocytopenia in brown trout, *Salmo trutta* L. General and Comparative Endocrinology 53:252–259.

Plumb. J. A. 1994. Health Maintenance of Cultured Fishes: Principal Microbial Diseases. CRC Press, Boca Raton, Florida.

Post, G. 1987. Textbook of Fish Health. TFH Publications, Neptune, New Jersey.

Ravndal, J., T. Løvold, H. B. Bentsen, K. K. Røed, T. Gjedrem, and K. Røvik. 1994. Serum iron levels in farmed Atlantic salmon: family variation and associations with disease resistance. Aquaculture 125:37–45.

Romalde, L., B. Magariños, F. Pazos, A. Silva, and A. E. Toranzo. 1994. Incidence of *Yersinia ruckeri* in two farms in Galicia (NW Spain) during a one-year period. Journal of Fish Diseases 17:533–539.

Salonius, K., and G. K. Iwama. 1993. Effects of early rearing environment on stress response, immune function, and disease resistance in juvenile coho (*Oncorhynchus kisutch*) and chinook salmon (*O. tshawytscha*). Canadian Journal of Fisheries and Aquatic Sciences 50:759–766.

Shuter, B. J. 1990. Population-level indicators of stress. American Fisheries Society Symposium 8:145–166.

Sindermann, C. J. 1988. Biological indicators and biological effects of estuarine/coastal pollution. Water Resources Bulletin 24:931–939.

Sindermann, C. J., and J. F. Chenoweth. 1993. The fungal pathogen *Ichthyophonus hoferi* in sea herring, *Clupea harengus*: A perspective from the western North Atlantic. International Council for the Exploration of the Sea, Copenhagen, Denmark.

Snieszko, S. F. 1954. Therapy of bacterial fish diseases. Transactions of the American Fisheries Society 83:313–330.

Soares, F., M. T. Dinis, and P. Pousao-Ferreira. 1994. Development of the swim bladder of cultured *Sparus aurata* L.: A histological study. Aquaculture and Fisheries Management 25:849–854.

Summerfelt, R. C. 1991. Symposium on Fish and Crustacean Larviculture. Special Publication No. 15, European Aquaculture Society, Ghent, Belgium.

Swain, D. P., and B. E. Riddle. 1990. Variation in agonistic behavior between newly emerged juveniles from hatchery and wild populations of coho salmon, *Oncorhynchus kisutch*. Canadian Journal of Fisheries and Aquatic Science 47:566–571.

Thompson, I., A. White, T. C. Fletcher, D. F. Houlihan, and C. J. Secombes. 1993. The effect of stress on the immune response of Atlantic salmon (*Salmo salar*) fed diets containing different amounts of vitamin C. Aquaculture 114:1–18.

Thosen, J. C. (editor). 1994. Procedures for the Detection and Identification of Certain Fish Pathogens, 4th ed. Fish Health Section, American Fisheries Society, Bethesda, Maryland.

Tucker, C. S., and E. H. Robinson. 1990. Channel Catfish Farming Handbook. Van Nostrand Reinhold, New York.

Wald, L., and M. A. Wilzbach. 1992. Interactions between native brook trout and hatchery brown trout: Effects on habitat use, feeding, and growth. Transactions of the American Fisheries Society 121:287–296.

Wallace, J. C., and A. G. Kolbeinshavn. 1988. The effect of size grading and subsequent growth in fingerling Arctic charr, *Salvelinus alpinus* (L.). Aquaculture 73:97–100.

Walters, G. R., and J. A. Plumb. 1980. Environmental stress and bacterial infection in channel catfish, *Ictalurus punctatus* Rafinesque. Journal of Fish Biology 17:177–185.

Wedemeyer, G. A., and C. P. Goodyear. 1984. Diseases caused by environmental stressors. Pages 424–434 *in* O. Kinne (ed.), Diseases of Marine Animals, vol. 4, part 1: Introduction, Pisces. Biologische Anstalt Helgoland, Hamburg, FRG.

Wedemeyer, G. A., F. P. Meyer, and L. Smith. 1976. Environmental Stress and Fish Diseases. TFH Publications, Neptune, New Jersey.

Wise, D. J., T. E. Schwedler, and D. L. Otis. 1993. Effects of stress on susceptibility of naive channel catfish in immersion challenge with *Edwardsiella ictaluri*. Journal of Aquatic Animal Health 5:92–97.

6
Managing Pathogen Exposure

INTRODUCTION

As discussed in previous chapters, maintaining a fish—pathogen–environment relationship that is favorable to the fish is critical to the continuing success of any hatchery, fish farm, or aquaculture facility. The most important method of achieving the required favorable relationship is to manage the interactions of fish with biological, chemical, and physical conditions in the rearing environment so as to minimize the effects of stress on resistance to infectious and noninfectious diseases. In some cases, however, it is practical to improve the fish—pathogen–environment relationship by reducing or eliminating the pathogen load of the water supply as well. At present, the most promising approaches are quarantine, broodstock segregation, and other biological control methods that exclude pathogens from the hatchery facility initially; and water treatment systems using ultraviolet light (UV), chlorine, or ozone as disinfectants.

BIOLOGICAL METHODS

Preventing the initial introduction of facultative and obligate pathogens to aquaculture facilities should always receive a high priority when developing methods to improve fish health. Practical procedures include egg disinfection, quarantine of incoming fish, segregation of broodstock carriers, and the use of groundwater supplies (wells or springs) where available. In some cases, biological methods can also be used to reduce the obligate and facultative pathogen load in surface water supplies. If the hatchery is supplied by a small stream, resident fish populations that are reservoirs of obligate pathogens can sometimes be removed. It may also be possible to use fish screens to keep resident fishes from migrating upstream of the hatchery intakes during the rearing season. If groundwater can be used, facultative pathogens will usually be absent and removing resident fishes and other aquatic animal life will eliminate obligate pathogens as well. However, biologists should keep in mind that groundwater will unavoidably be repopulated by a variety of facultative microorganisms soon after it is exposed to the air in headboxes, raceways, or ponds. In rare cases, groundwater may come to the surface already contaminated with microorganisms if fractured rock areas or limestone channels exist.

WATER TREATMENT SYSTEMS

As fish culture continues to become more intensive worldwide, surface water supplies, many of which contain infected wild fish populations, are having to be increasingly utilized in lieu of the more desirable springs or wells. Hatcheries may also be obliged to recirculate water to compensate for inadequate single-pass supplies. In both situations, it may become imperative to control diseases transmitted by water-borne facultative and obligate pathogens by disinfecting the incoming or recirculated water with ultraviolet light or with chemicals such as ozone. For economic reasons, disinfection with ultraviolet (UV) light is usually the treatment of choice, however chemical disinfectants such as ozone, chlorine, or iodine may also be cost effective depending on the situation. Of the chemical disinfectants, ozone and chlorine are the most widely used, although iodine and other oxidizing chemicals such as potassium permanganate and hydrogen peroxide have all been occasionally employed. For example, adding elemental iodine to a hatchery water supply at 0.1 mg/L inactivated infectious hematopoietic necrosis virus and prevented its horizontal transmission (Batts et al. 1991).

Theoretically, hatchery water treatment systems could be operated to achieve either sterilization or disinfection. Sterilization, the total destruction of all living organisms in the water supply, is not necessary and is too expensive to be practical in any case. Disinfection, the destruction of organisms harmful to the user of the water, normally gives an entirely adequate level of fish pathogen control (Torgersen and Håstein 1995). In practice, disinfection usually also destroys some microorganisms not harmful to the fish such as algae, zooplankton, or nitrifying bacteria. However, inactivating such potentially helpful microorganisms is rarely detrimental unless it occurs on a large scale.

Regardless of the type of disinfectant employed, the rate at which it kills at a particular concentration is theoretically proportional to the exposure time t, and the number of microorganisms initially present. Because this relationship was first expressed mathematically by Chick (1908), it has become known as Chick's law of disinfection:

$$\log_e (N_t/N_0) = -kt \tag{6.1}$$

where

N_0 = number of viable microorganisms initially present,
N_t = number of microorganisms remaining viable at time t,
k = proportionality constant, and
t = exposure time (usually minutes).

As shown by equation 6.1, the fraction of pathogens remaining viable (N_t/N_0) decreases logarithmically while exposure time to the disinfecting agent increases arithmetically. However, the quantity $\log_e(N_t/N_0)$ never reaches zero. That is, the probability that all the pathogens will be killed never reaches certainty.

For convenience, pathogen kill data are usually expressed as a percentage kill or as a $\log_{10}$ reduction in numbers after a given exposure time. A 99.9% kill, for example, would be a 3-log reduction in viable pathogens. Expressing equation 6.1 in terms of base 10 logarithms, inverting N_t/N_0 to eliminate the negative sign, and rearranging to express the fraction killed as a function of time gives:

$$t = 2.3/k \log(N_0/N_t) \tag{6.2}$$

Calculating the contact time (t) to use for a particular disinfectant requires that a probability decision be made because the death of a microbial population is a statistical phenomenon. For example, one guideline for aquaculture equipment disinfection (nets, tanks) is a 99.999% kill of the viable pathogens present or a 5-log reduction. This standard allows only one chance in 100,000 of failure to kill the organisms in question. The contact time (t) required to achieve this margin of safety is calculated by setting N_0 equal to 1 and N_t equal to 0.00001. For water treatment, a 3-log reduction or a 99.9% kill is usually more than adequate. As mentioned, the probability of failing to kill all of the microorganisms becomes smaller as N_t becomes smaller but it never reaches unity. In practice, the rate of pathogen inactivation by disinfectants often deviates somewhat from the theoretical logarithmic rate described by Chick's law. This may be due to the microorganisms forming clumps, or because the disinfectant is reacting with other materials present in addition to reacting with the pathogen (Wei and Chang 1975).

In addition to contact time, the kill achieved by a particular disinfectant is also a function of its concentration. The relationship between contact time and disinfectant concentration is given by Watson's law (Watson 1908):

$$t = kC^n \tag{6.3}$$

where

t = contact time (usually minutes),
C = disinfectant concentration (usually mg/L),
n = coefficient of dilution, and
k = proportionality constant.

As shown by equation 6.3, decreasing the disinfectant concentration can cause an exponential increase in the time required to achieve a given kill—depending on the value of the coefficient of dilution (n). The value of n must be determined empirically because it varies with the microorganism, type of disinfectant, and environmental variables such as pH and temperature. Rearranging equation 6.3 and taking the logarithm of both sides results in the following equation:

$$\log_e t = k - n\log_e C \tag{6.4}$$

The value of the coefficient of dilution (n) is the slope of the line obtained by plotting $\log_e t$ against $\log_e C$. In practice, n is rarely calculated, but it is conceptually important. For disinfectants with n values <1, contact time is more important than concentration. Practically speaking, this means that the disinfectant will retain its effectiveness over a range of dilutions. When the value of n is near unity, contact time and concentration have approximately equal effects. If n is large (values >1), the disinfectant can be diluted only a small amount before it loses its practical effectiveness. As an example, the n value for chlorine is usually near 1 while mercurous chloride has an n of about 5. Diluting a chlorine solution by 50% will only double the contact time required, but halving the concentration of a mercurous chloride disinfectant solution will increase the contact time required by a factor of $2^5 = 32$.

Chlorination

Chlorine in gas, liquid, or solid form has been used to disinfect nets, tanks, and other hatchery equipment for many years. It is also commonly used to inactivate fish pathogens in hatchery or laboratory effluent discharges. When large volumes of water such as effluent discharge streams are to be disinfected, chlorine gas (Cl_2) is almost always the most economical form to use. For smaller volumes, such as hauling truck tanks and tanks for disinfecting nets, hypochlorite salts such as calcium or sodium hypochlorite are usually employed.

Chlorine gas is normally injected into water through a diffuser. Because of its high solubility (≈ 7 g/L at 20°C), gas transfer is highly efficient. In water of pH greater than about 3, and at the concentrations of chlorine normally used in fisheries applications (<500 mg/L), virtually all the dissolved molecular chlorine will immediately hydrolyze into hypochlorous acid, which is also strongly germicidal.

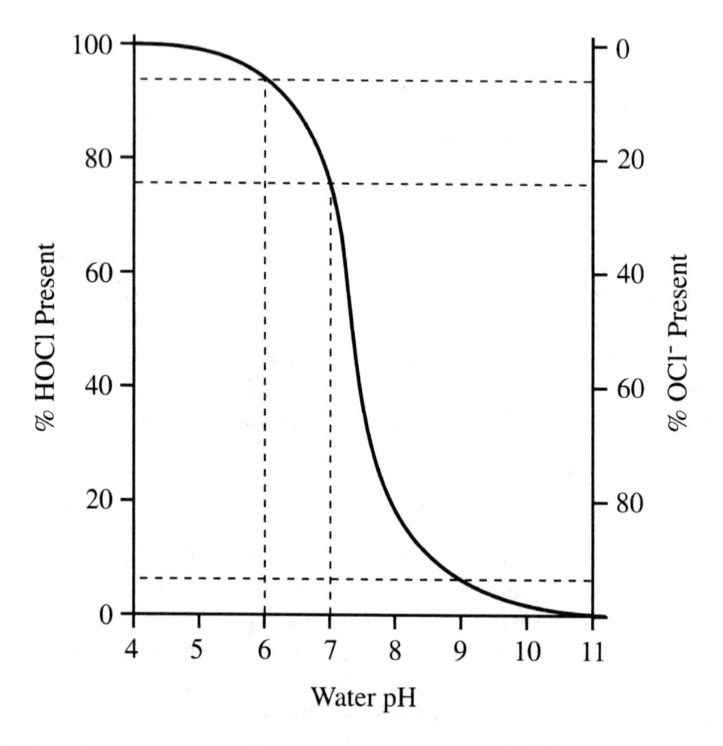

Figure 6.1. The effect of pH on the proportions of strongly germicidal hypochlorous acid and the less germicidal OCl^- present in chlorine disinfectant solutions.

$$Cl_2 + H_2O \rightleftarrows HOCl + H^+ + Cl^-$$

Hypochlorous acid, however, dissociates to form the hypochlorite ion OCl^- which has less germicidal activity.

$$HOCl \rightleftarrows OCl^- + H^+$$

The relative proportions of HOCl and the less germicidal OCl^- that will be present in a disinfectant solution depends on the pH. More OCl^- is formed in alkaline (hard) water and more HOCl in acidic (soft) water. As shown in Figure 6.1, about 95% of the HOCl is dissociated into the less germicidal OCl^- at pH 9, whereas at pH 6, about 95% of the added chlorine will be in the more effective HOCl form.

For disinfecting equipment such as boots, nets,[25] and tanks, chlorine is usually added to the water as one of its hypochlorite salts (usually sodium or calcium

25. A chlorine concentration of 50 mg/L is recommended for nets to avoid damaging them.

hypochlorite) to achieve a final concentration of about 100 mg/L. These salts are highly soluble and upon dissolving immediately ionize to form hypochlorite ions.

$$Ca(OCl)_2 \rightleftarrows Ca^{2+} + 2OCl^-$$

Depending on the pH, the hypochlorite ions will remain as OCl^- or be converted to the more strongly germicidal hypochlorous acid as just discussed. For this reason, acetic acid (or other weak acid) is sometimes added to chlorine disinfectant solutions in hard water areas to lower the pH to about 6. This "activates" the solution by promoting the formation of HOCl.

The HOCl and OCl^- formed when Cl_2 or hypochlorite salts are added to water account for most of the germicidal activity of chlorine solutions. The sum of the concentrations of HOCl and OCl^- is termed the *free chlorine residual*. However, HOCl and OCl^- (and Cl_2 itself) are strong oxidizing agents that will react with other compounds present in the water in addition to inactivating microorganisms. In fish culture systems, variable concentrations of ammonia and other nitrogenous wastes are almost always present and a variety of chloramines, including monochloramine, dichloramine, and trichloramine are invariably produced:

$$NH_3 + HOCl \rightleftarrows NH_2Cl + H_2O$$
$$NH_3 + 2HOCl \rightleftarrows NHCl_2 + 2H_2O$$
$$NH_3 + 3HOCl \rightleftarrows NCl_3 + 3H_2O$$

Ironically, chloramines usually are weaker germicides than chlorine itself but are usually much more toxic to fish. They are also not as easily removed from water by aeration, sunlight, or other dechlorination methods as free chlorine is and thus tend to persist for long periods in the aquatic environment. The total chlorine residual formed when chlorine is added to water thus usually consists of both HOCl and OCl^- (the free chlorine residual) and chloramine compounds, all of which have differing germicidal activity and fish toxicity. The total concentration of the chloramine compounds formed when chorine is added to water is termed the *combined chlorine residual*.

Because of these and other reactions, the residual that develops when chlorine is added to water builds up in a characteristic manner historically termed the *breakpoint chlorination curve* (Figure 6.2). As chlorine is first added, it oxidizes any reducing agents present, such as nitrites and ferrous ions. The chloride ions (Cl^-) formed in the process have no germicidal activity and no residual develops. As more chlorine is added, it then begins to react with any ammonia or other nitrogenous compounds present forming chloramines. These generally have weak germicidal activity and a residual (the combined chlorine residual) begins to accumulate (Point A in Figure 6.2). As still more chlorine is added, it begins to oxi-

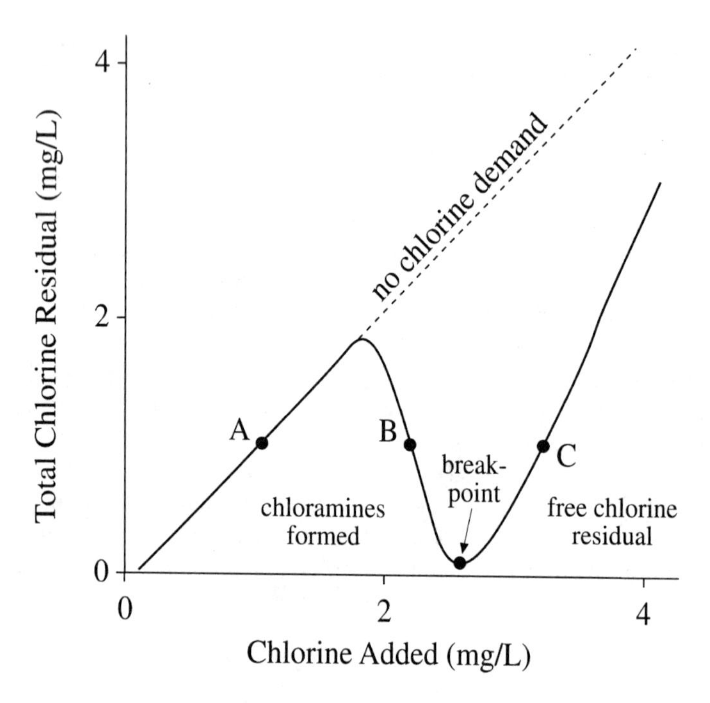

Figure 6.2. As chlorine is progressively added to most water supplies, the residual builds up in a characteristic manner termed the *breakpoint chlorination curve*. Because of chloramine formation, the germicidal activity at points A, B, and C along the curve can be quite different, although the total chlorine concentration is about the same. The dotted line shows how a residual builds up under ideal conditions in water with no chlorine demand.

dize the chloramines, and compounds such as nitrogen trichloramine (NCl_3) and nitrous oxide are produced. The combined chlorine residual then falls (Point B in Figure 6.2). Adding more chlorine completes the oxidation of the chloramines and a residual again begins to accumulate. After this point, historically referred to as the breakpoint, a residual with considerable germicidal activity (HOCl and OCl^-) begins to accumulate (Point C in Figure 6.2). The practical importance of the breakpoint chlorination curve is that the total residual chlorine concentration can be the same at several points along the curve while the germicidal activity is not. For example, the total chlorine concentrations at points A, B, and C in Figure 6.2 are all about 1.5 mg/L but might have killing times for bacterial spores of 90, 90, and 3 minutes respectively. For this reason, water chemistry analyses for chlorine should always include a free residual as well as a total residual determination (discussed in Chapter 3).

In the absence of chloramines, the bactericidal, virucidal, and fish toxicity effects of chlorine are due primarily to HOCl and, to a lesser extent, OCl^-. Hypochlorous acid is a strong oxidant and destroys both microorganisms and gill tissue by causing nonspecific damage to proteins and other constituents of cell walls and cell membranes. It also passes easily into cells where it oxidizes enzymes and the purine and pyrimidine ring structure of DNA and RNA. The OCl^- ion has about the same molecular weight as HOCl but its electrical charge impedes penetration through cell membranes making it a weaker disinfectant and also less toxic to fish.

The efficacy of chlorine disinfection is affected by turbidity as well as by pH. Suspended particulates such as clay or organic matter tend to protect microorganisms by shielding them from direct chlorine exposure or by reacting with it. However, water quality conditions such as dissolved minerals generally have little effect. Added NaCl increases the susceptibility of some viruses to chlorine inactivation by causing conformational changes in the capsid (outer protein shell) that facilitate HOCl penetration (Sharp and Leong 1980). This ionic strength effect has important implications for experimental work but is probably of little practical importance in hatchery disinfection.

Chlorination to destroy fish pathogens in effluents from fish holding facilities (usually laboratories or hatcheries) requires a free residual of 1–3 mg/L with a contact time of 10–15 minutes. If exotic pathogens are involved, higher concentrations (3–5 mg/L) with a contact time of 15–30 minutes should be used to ensure an ample margin of safety.

For equipment and tank disinfection, a chlorine concentration of 100–200 mg/L with a contact time of at least 30 minutes is usually recommended. For convenience, the chlorine is usually applied as a NaOCl solution or as granular $Ca(OCl)_2$. Of necessity, disinfection protocols will vary somewhat but for fish transport tanks, sufficient water to cover the intakes of the recirculating pumps or spray agitators is poured in and the calculated amount of liquid or dry chlorine added (e.g., 15 g HTH, 70% available chlorine, per 100 L). If the water is alkaline, sufficient acetic acid to decrease the pH to about 6 should be added to activate the chlorine[26]. The solution is pumped through the system for at least 30 min. Air or oxygen must be flowing through the diffusers during this time. A summary of chlorine dose and contact times generally recommended for fishery applications is given in Table 6.1. As mentioned, acetic acid should be used to adjust the pH to below 7 in hard (alkaline) water areas. Also, in view of recent data showing that bacteria can become resistant to chlorine (Jeffrey 1995), these dose and contact times may need to be increased at some hatcheries.

After disinfection, tanks, nets, and other equipment must be thoroughly rinsed with chlorine-free water. Residual chlorine can also be neutralized with a sodium sulfite or thiosulfate rinse before the equipment is used to handle fish. The concentration of sodium thiosulfate needed is 7.4 ppm per ppm chlorine to be neutralized. For a 100 mg/L chlorine solution, this would mean adding sufficient

26. Safety note: never add acids directly to chlorine compounds; toxic Cl_2 gas will be liberated.

TABLE 6.1. Minimum recommended chlorine dose and contact times to inactivate common fish pathogens and disinfect hatchery equipment and effluent discharges.

Pathogen/Application	Dose (mg/L)	Contact Time	Comments
	Pathogen Inactivation		
Aeromonas salmonicida	0.1	30 sec	99.9% kill
Renibacterium salmoninarum	0.05	18 sec	99% kill
Yersinia ruckeri	0.05	30 sec	99.9% kill
IPNV (*infectious pancreatic necrosis virus*)	0.2	10 min	Soft water
IPNV	0.7	2 min	Hard water
IHNV (*infectious hematopoietic virus*)	0.5	5 min	Soft water
IHNV	0.5	10 min	Hard water
	Fishery Application		
Nets, boots	50[a]	5 min	Neutralize after use
Tanks	100	30 min	Neutralize after use
Dried out ponds	200	Allow to dissipate naturally	
Hatchery wastewater pathogens	0.3–0.5	10 min	Absent exotic

Source: Compiled from Wedemeyer et al. (1979); Pascho et al. (1995); and the experience of the author.

[a]Higher concentrations can damage nets.

sodium thiosulfate to achieve a concentration of 740 mg/L, or about 5 g of sodium thiosulfate pentahydrate per gallon (Jensen 1989). As a precaution, commercially available chlorine test paper or a chemical method such as the *o*-tolidine determination can be used to confirm that chlorine residuals have dissipated to safe levels or have been adequately neutralized. Allowing a chlorine residual on nets and other equipment to dissipate naturally over a period of several days is often done but it is unreliable and exposes the fish to some degree of risk.

Removing chlorine from water is also sometimes necessary in fisheries work. Although chlorination is an effective method for destroying pathogens in hatchery water supplies, it poses a serious physiological risk to the fish. Even with fail-safe dechlorination systems, risk of fish toxicity remains and continuous monitoring with an automated chlorine analyzer is highly recommended. In practice, chlorination is normally only used to inactivate pathogens in hatchery (or laboratory) discharge streams. However, most species in receiving waters are also highly sensitive to chlorine and any chloramines produced are even more toxic. Thus, success in using chlorine for pathogen control in waste discharge streams also requires the virtual elimination of chlorine and chloramines before the effluent reaches the receiving water. Other applications include dechlorinating municipal water so it can be used for fish rearing, and neutralizing chlorine residues on equipment that has been previously disinfected so it can be used immediately.

Dechlorination methods must remove both chlorine and chloramines to be useful for fisheries work. Photochemical decomposition by UV irradiation and treatment with ozone have considerable promise but high cost has limited the application of these methods. Small concentrations of chlorine (e.g., 0.05 mg/L) can be inexpensively removed by aeration, but ample ventilation is required or the chlorine will simply redissolve. Aeration will not remove chloramines. In most cases, either chemical reducing agents or adsorption techniques are the only practical alternatives. Adsorption by activated charcoal is effective in eliminating chlorine and chloramines down to the 10–20 μg/L range. Chemical reducing agents such as sodium thiosulfate must be added to remove the residual toxicity.

Of the chemical and absorption methods, the two most commonly used are treatment with sulfur compounds (sulfur dioxide gas, sodium thiosulfate, sodium sulfite), and filtration through activated carbon.

Dechlorination with sulfur dioxide (SO_2) gas is inexpensive, easy to control, and the required equipment is commercially available. If necessary, gas chlorination equipment can also be easily modified to inject SO_2 instead of Cl_2. SO_2 reacts with water to form sulfite ions (SO_3^{-2}), which then reduce the chlorine and chloramines to chloride ions and inconsequential amounts of acids in a 1:1 stoichiometric ratio. These reactions are so rapid that contact time is not an important consideration.

$$SO_2 + H_2O \rightleftarrows H_2SO_3$$
$$HOCl + SO_3^{-2} \rightleftarrows H_2SO_4 + HCl$$
$$RNHCl + SO_3^{-2} \rightleftarrows RNH_2 + H_2SO_4 + HCl$$

When smaller volumes of water are involved, sodium sulfite (Na_2SO_3) or thiosulfate ($Na_2S_20_3$) solutions are convenient and inexpensive to use for dechlorination. These compounds hydrolyze to produce sulfite ions which then react with chlorine in the same way as SO_2. Theoretically, a concentration of 7.4 mg/L of sodium thiosulfate pentahydrate solution will neutralize 1 ppm of chlorine. However, a concentration of 8 mg/L is recommended because it provides a small margin of safety. If sodium sulfite is used, the stoichiometric neutralization ratio is about 6:1. Again, a higher treatment concentration is recommended to provide a safety margin. In addition, sulfite solutions are unstable in storage and should be used within a day or two of preparation.

For facility wastewater discharges in the 50–100 gallons per minute (gpm) range, the cost of sodium thiosulfate treatment may be about the same as the SO_2

needed for gas dechlorination, and safety concerns are greatly reduced. Equipment needs are also simpler—a solution reservoir tank and a metering pump.

Filtration through activated charcoal (carbon) is perhaps the most commonly used dechlorination method in fisheries work. Activated carbon (C^*) chemically reacts with both chlorine and chloramines, converting them into innocuous amounts of carbon dioxide and ammonium salts.

$$C^* + HOCl \rightleftarrows CO_2 + HCl$$
$$C^* + NH_2Cl + H_2O \rightleftarrows CO_2 + NH_4Cl$$

Fresh activated charcoal will reliably reduce the free and bound chlorine concentration down to a few μg/L but 100% removal almost always requires supplemental sulfite or thiosulfite injection, especially as the filters age (Seegert and Brooks 1978). In the experience of the author, a flow-through filter containing 12 ft^3 of granular activated carbon will reliably dechlorinate about 25 gpm of water containing a chlorine residual of 0.05 mg/L for a period of about one year before needing replacement.

The fish toxicity of chlorine is high. At the concentrations of 0.1–0.3 ppm usually found in drinking water supplies, chlorine will kill most commercially important species within minutes at any pH. As mentioned, the OCl^- form is less toxic than HOCl but the practical importance of this is slight because little OCl^- is present over the pH range 5–9 typical of most fish rearing conditions. Environmental conditions such as temperature and dissolved oxygen (DO) also influence chlorine toxicity but again, the practical importance of any protection gained is slight. Maintaining chronic exposure levels at <3–5 μg/L will protect both the cold- and warmwater fishes used in aquaculture from gill damage. Most fishes can tolerate exposures as high as 0.05 ppm for short periods (up to 30 min), because the gill epithelium can recover from the damage that occurs.

Ultraviolet Light

The germicidal properties of sunlight have been recognized since the time of Louis Pasteur but it was not until 1893 that the ultraviolet (UV) portion of the solar spectrum was identified as actually responsible (Ward 1893). The term "ultraviolet light" is the misnomer applied to that portion of the sun's invisible electromagnetic radiation with wavelengths shorter than those of visible light but longer than X-radiation. As illustrated in Figure 6.3, the UV band begins at about 400 nm, the shortest wavelength of visible light (violet), and extends down to about 100 nm; the lower end of the X-radiation band. By convention, the UV radiation spectrum is further subdivided into the UV-A band, radiation between 400 and 315 nm; the UV-B band (315–290 nm); and the UV-C band, (290–200 nm, the most dangerous to living organisms). The UV-A band (400–315 nm) is the portion

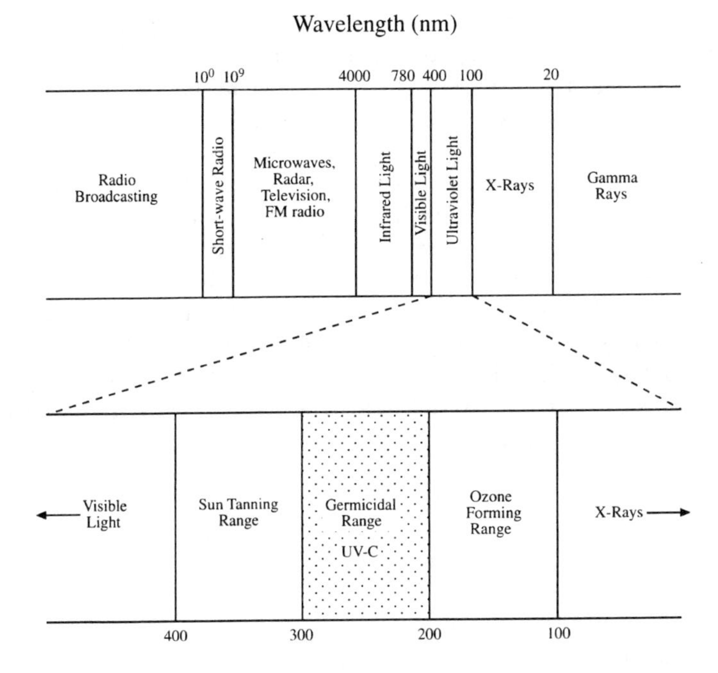

Figure 6.3. Diagrammatic illustration showing the place of ultraviolet light (UV) in the total spectrum of electromagnetic radiation emitted by the sun. Most of the killing activity of UV radiation is due to wavelengths near 260 nm in the UV-C band.

largely responsible for melanocyte activation in skin tissue (tanning) while radiation in the UV-B band stimulates Vitamin D production in the skin.

As mentioned, sunlight includes significant amounts of UV radiation and thus has germicidal action, provided the "light" has not passed through layers of glass, water, or plastics that absorb it[27]. Historically, the earth's ozone layer has blocked essentially all of the sun's UV radiation between 200–290 nm (UV-C band, the most dangerous to living organisms), and most of the radiation between 290–315 nm (UV-B band). However, biologically significant amounts of UV can still reach the earth's surface. The intensity of the total UV radiation at ground level can be as high as 150 microwatts per cm^2 ($\mu W/cm^2$) on clear days in the northern latitudes of the United States decreasing to essentially zero at night (Kerr and McElroy 1993). Thus, sunlight is of some use in disinfecting equipment such as nets, and ponds that have been dried out. However, atmospheric conditions are too variable to make it completely reliable.

27. Mylar plastic films 0.25 mm thick or greater will block most solar UV at wavelengths shorter than 315 nm.

The killing effect of UV light is strongly dependent on the wavelengths used. Electromagnetic radiation, whether X-rays, ultraviolet, visible light, or radio waves, occurs in discrete energy units termed *photons*. Photons of shorter wavelength radiation possess greater energy and can inflict more biological damage than photons of longer wavelength radiation. That is, as the wavelength decreases, the energy per photon increases. Photons of electromagnetic radiation in the UV wavelength band are energetic enough to cause considerably more biological damage than the radiation visible as light, but considerably less damage than X-radiation. The most lethal UV wavelengths are those between 255–265 nm, the narrow band at which DNA, RNA, and other nucleic acids most strongly absorb UV radiation and are disrupted by it (Figure 6.4). Nucleic acid absorption peaks at 260 nm. Amino acids containing aromatic ring structures absorb UV at 280 nm causing damage to proteins, enzymes and other cell constituents, which also contributes to the killing effect.

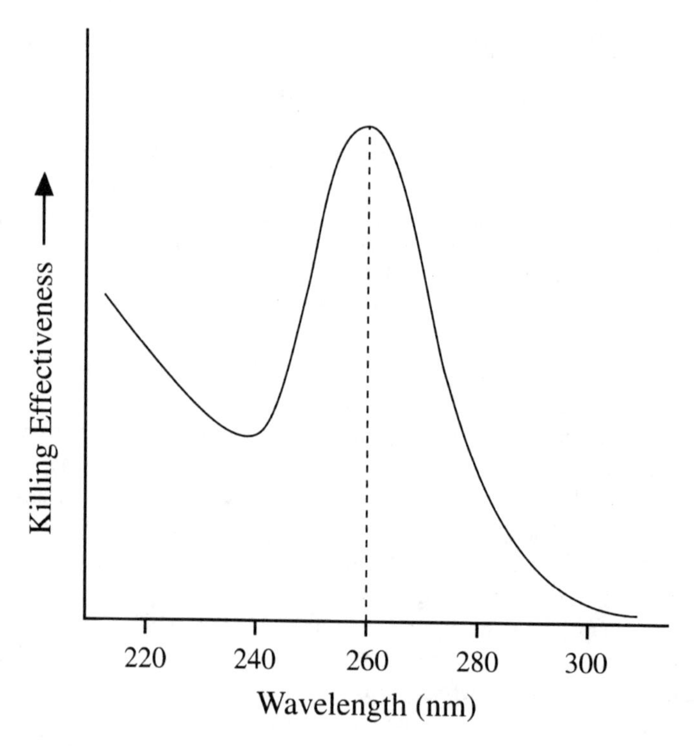

Figure 6.4. The killing effectiveness of UV light depends on its wavelength. Nucleic acids absorb strongly at 260 nm causing lethal damage to DNA/RNA. Amino acids containing aromatic ring structures absorb UV radiation at 280 nm causing damage to peptides, structural proteins, and enzymes. UV "light" radiation at these wavelengths is invisible and can cause severe damage to the cornea and retina. Eye protection is mandatory.

In common with other disinfectants, the biological damage caused by UV light is proportional to its intensity and to the exposure time. In contrast however, temperature has little effect on its biocidal activity.

It is important to note that most microorganisms possess biochemical mechanisms enabling them to repair UV radiation damage. As with chemical disinfectants, the same killing effect can be achieved by short exposure times and high intensities or by low intensities and longer exposure times—up to a certain fraction of the DNA repair time of the target organism. Of the nucleic acid repair mechanisms, photoreactivation is of particular importance. If UV damaged pathogens are allowed access to normal light for even a few hours, much of the DNA/RNA damage can be repaired by light-activated enzyme systems. The practical result is a substantial increase in the number of viable organisms surviving the exposure (as if a much lower UV dose had been used).

UV Treatment Systems

UV water treatment systems offer continuous control of both obligate and facultative pathogens in hatchery water supplies. In facilities using recirculated water, some degree of control over pathogens transmitted horizontally (fish-to-fish) may be obtained as well. Diseases transmitted only vertically are usually beyond the reach of water treatment systems. In any case, to inactivate microorganisms important to aquaculture, UV radiation must penetrate water. Fortunately, water absorbs light less strongly as the wavelength becomes shorter (blue end of the visible spectrum) and is relatively transparent to UV radiation at wavelengths shorter than about 250 nm. However, turbidity and dissolved mineral salts reduce transmission significantly. Thus, seawater absorbs UV more strongly than most freshwaters because of its higher salinity. In fish culture situations, turbidity from plankton, silt, and suspended organic matter is usually the most serious cause of reduced UV transmission. Sand prefilters should always be used unless the natural turbidity is <20 Natelson Turbidity Units (NTUs). Even under ideal conditions however, inactivating pathogens in water requires 5 to 10 times the UV dosage needed for microorganisms suspended in the air.

To deliver the high dose rates needed, lamps are required that produce substantial amounts of UV radiation at wavelengths near 260 nm. Small amounts of UV are produced by all lamps that produce visible light. For example, standard incandescent lamps, especially those of 500 watts or above, produce measurable amounts of UV in the germicidal range. However, the amount of UV produced per watt of power consumed is too small to be practical for fish disease control. The UV output of standard fluorescent lamps is high enough to kill developing salmonid eggs at a distance of 40 cm (Dey and Damkaer 1991), but still not adequate for water disinfection. The amount of UV radiation can be increased into the practical range by omitting the phosphor coating that produces visible light and adding an additional amount of mercury. The lamp filament vaporizes the mercury and the electrical arc produced excites its outer electrons into orbitals at higher energy levels. As the electrons fall back to their ground state, they emit most of the previously absorbed energy as radiation in the UV range of wavelengths (200–400 nm).

Both high and low pressure mercury vapor lamps have been developed for water treatment systems. For fish pathogen inactivation, low pressure lamps are the most widely used because they are relatively inexpensive and they emit about 95% of their UV radiation in a narrow band centered around 254 nm, very close to the 260 nm wavelength producing the most effective killing action. Low-pressure mercury vapor lamps are available as hot cathode, cold cathode, and high-intensity germicidal lamps used for disinfecting air. Hot cathode lamps operate at standard line voltage using a ballast similar to a fluorescent lamp. The electrodes are metal oxide coated tungsten filaments located in each end. Lamp life is a function of filament life (reduced by frequent on and off switching), and solarization of the lamp glass. Solarization is the slow darkening of glass caused by prolonged exposure to UV. It reduces lamp output by 3–4% per 1,000 hours of use. The lamp is considered to have reached the end of its useful life when output has declined to 60% of its original level.

Cold cathode lamps contain argon and neon in addition to the mercury vapor and require much higher operating voltages. Because there are no filaments to burn out, lamp life is determined primarily by the rate of solarization. They produce about the same amount of UV radiation per watt of electricity used as hot cathode lamps. Unfortunately, cold cathode lamps also radiate significant amounts of UV at wavelengths (100–200 nm) that generate ozone and hydrogen peroxide (H_2O_2) in the air and water. The fish toxicity of both these compounds is high. Fortunately, ozone and H_2O_2 production can be minimized by adjusting the UV transmission characteristics of the glass used to make the tube.

Physically, UV units for fish culture applications usually consist of low pressure, hot cathode lamps submerged in the water to be treated. Typically, the lamps are mounted radially in the center of a cylindrical chamber through which the water flows (Figure 6.5). Individual lamps are usually enclosed in a sleeve made of UV-transmitting glass or quartz. This keeps the water from contacting the lamp but allows the UV radiation to pass into the water and inactivate the target microorganisms. Although the quartz sleeve can be eliminated, it provides several advantages. Water temperature has less influence on lamp output and replacement is much easier because a watertight seal around the lamp is not needed. Also, the water pressure in the chamber can be higher because it is limited by the strength of the quartz tube and its seal, not by the physical strength of the UV lamp.

Because the quartz tube is in contact with the water, it will eventually become coated with mineral deposits, sediment particles, and other types of fouling. Unless these deposits are periodically removed, they will increasingly block the UV radiation. Tube cleaning is thus an indispensable part of system maintenance. In

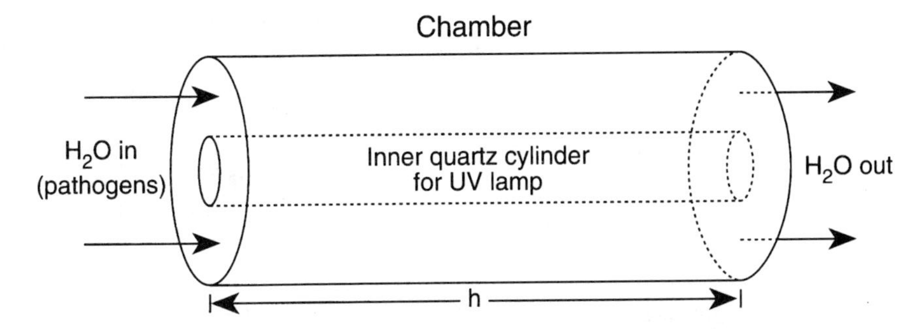

Figure 6.5. Diagram of a UV water treatment chamber showing parameters used in calculating flow rates (contact time) needed to ensure that suspended pathogens receive the lethal dose of 30,000 μwatt·sec/cm^2.

systems with UV sensors, cleaning can be done when the intensity drops to a predetermined level. Otherwise, cleaning should be scheduled at regular intervals based on experience. Some systems employ a wiper block arranged such that the flow of water through the unit drives it back and forth over the quartz sleeve. Automatic cleaning systems reduce labor requirements and avoid the necessity of monitoring UV intensity to ensure that manual cleaning has been performed faithfully.

The UV dose rate (intensity and contact time) required for fish disease control is expressed as the dose of ultraviolet radiation in microwatts (μW) per cm^2 of area irradiated per second of contact time (μW·sec/cm^2). Dosages required to inactivate fish pathogens vary somewhat but most of the common bacteria and viruses are easily destroyed by 30,000 μW·sec/cm^2 or less. Larger parasites such as *Icthyopthirius* tomites (Ich), and the protozoan parasite *Ichthyobodo (Costia) necatrix* can survive UV dosages as high as 300,000 μW·sec/cm^2 (Caufield 1991). For fish disease control, the accepted guideline is a dose rate of 30,000 μW·sec/cm^2 at the end of useful lamp life (7500 hours), or 55,000 μW·sec/cm^2 after 100 hours. This dosage will achieve the 3-log reduction in pathogen numbers (99.9% kill) that is the presently accepted standard for fish disease control. UV dose rates required for the inactivation of individual fish pathogens important in aquaculture may be considerably lower (or higher). A summary is given in Table 6.2.

Calculating the water flow through a UV unit that will result in suspended pathogens receiving the desired dose of 30,000 μW·sec/cm^2 is done using the lamp intensity and the effective volume of the chamber. Commercially available lamps typically produce UV at intensities of about 9,000 μW/cm^2. The effective chamber volume varies with the UV unit.

Example: Calculate the water flow through a cylindrical chamber with a standard UV tube mounted in its center that will result in a dose rate of 30,000 μW·sec/cm^2 to suspended microorganisms.

TABLE 6.2. Minimum ultraviolet light (UV) dosages recommended to inactivate common fish pathogens and disinfect hatchery water

Pathogen/Application	UV Dose Rate (μW·sec/cm^2)
Pathogen Inactivation	
Bacteria (99.9% inactivation)	
Aeromonas hydrophila	5,000
Aeromonas salmonicida	4,000
Vibrio anguillarum	4,000
Pseudomonas fluorescens	5,000
Fungus (inhibit the growth of hyphae)	
Saprolegnia sp.	230,000
Virus (99% decrease in infectivity)	
IPNV (*infectious pancreatic necrosis*)	150,000
IHNV (*infectious hematopoietic necrosis*)	2,000
CCV (*channel catfish virus*)	2,000
Herpesvirus salmonis	2,000
Parasites (99% decrease in infectivity)	
Myxobolus cerebralis	27,600
Fishery Application	
Recirculated water	30,000
Hatchery wastewater	30,000

Source: Compiled from Hoffman (1975); Yoshimizu et al. (1990).
Note: The very high dose rates required to inactivate IPN and *Saprolegnia* makes ozone an attractive alternative.

First, the effective chamber volume (ECV) must be calculated. Using h_c and h_q to represent the length of the chamber and quartz tube respectively (see Figure 6.5), the ECV is determined from:

$$ECV = V_{(chamber)} — V_{(quartz\ sleeve)} = \pi r^2 h_c - \pi r^2 h_q$$

The exposure time (usually seconds) needed for a pathogen suspended in water flowing through the chamber to accumulate a dose of 30,000 μW·sec/cm^2 can be determined from the following relationship:

$$ET\ (sec) = \text{Dose rate}\ (\mu W \cdot sec/cm^2) \div \text{Lamp intensity}\ (\mu W/cm^2)$$
$$ET = 30{,}000\ \mu W \cdot sec/cm^2 \div 9{,}000\ \mu W/cm^2 = 3.3\ sec$$

The water flow rate through the chamber that will provide this exposure time is given by:

$$WFR = ECV/ET$$

where

WFR = water flow rate (L/sec),
ECV = effective chamber volume (L), and
ET = exposure time (sec).

For this example, assume that the effective chamber volume = 3.3 L. The maximum water flow rate that can be used to ensure that suspended pathogens are exposed to UV for 3.3 seconds is:

$$WFR = ECV/ET = 3.3\ L/3.3\ sec = 1.0\ L/sec$$

That is, a UV unit with an effective chamber volume of 3.3 L using a 9,000 $\mu W/cm^2$ lamp can treat a maximum water flow[28] of 1.0 L/sec (or about 16 gallons per min).

Ozone

Ozone has long been recognized to have germicidal properties and has been used to disinfect municipal water supplies in Europe since the early 1900s. In the United States, however, ozone has been little used for this purpose. First, surface water supplies here have been less intensively utilized making problems with chloramines less severe and chlorination more economical. Second, ozone has a relatively short half-life (10–20 min) which makes it difficult to maintain a residual in water distribution systems. In contrast, ozone has long been used worldwide for removal of color, odor, and turbidity from recirculated water in commercial aquaria, from domestic drinking water supplies and from industrial and sewage treatment plant waste discharges. In contrast, the fisheries applications of ozone have centered around water disinfection. During the last two decades, ozone has been increasingly used to disinfect hatchery intake and effluent water streams, and to control pathogens, oxidize nitrite, remove color and odor, and reduce organic loadings in recirculated hatchery and commercial aquarium water supplies (Rosenthal and Wilson 1987). In shellfish aquaculture, ozone has also been used for the depuration of coliform-contaminated oysters and to inactivate the toxins

28. This example is based on clear water of <20 NTU turbidity. Also, lamp output progressively decreases as solarization occurs. For this reason, lamps are specified to deliver their rated output (e.g. 9,000 $\mu W/cm^2$) at the end of their useful life. At the start of service, their output is substantially higher.

produced by red tide dinoflagellates such as *Gymnodium breve*, *Gonyaulax catenella*, and *Gonyaulax tamarensis*, which are responsible for paralytic shellfish poisoning (Rosenthal and Wilson 1987).

Ozone is easily generated from air using either a high voltage electrical discharge or ultraviolet irradiation in the 100–200 nm wavelength band. In either method, the added energy disrupts some O_2 molecules producing oxygen atoms. When the newly produced oxygen atoms then collide most simply recombine into O_2, but some of them react to form molecules of unstable triatomic oxygen (O_3) termed *ozone*.

$$3O_2 + h\nu \rightarrow 2O_3$$

The quantity $h\nu$ represents the quantum unit of ultraviolet or electrical energy required to cause the reaction. Because of the high N_2 content of air, small amounts of toxic nitrogen oxides (NO_x) are also produced as by-products[29] but their significance in finfish aquaculture is inconsequential. Using oxygen-enriched air, or pure oxygen, as the feed gas will reduce or eliminate NO_x production and also increase the amount of ozone produced. As mentioned, ozone molecules in air are thermodynamically unstable and will spontaneously decompose back into oxygen over a period of several hours.

The ozone generators used in aquaculture are mostly electrical discharge units because of the large volumes of water to be treated, although UV ozone generators have occasionally been used. Basically, electrical discharge units consist of a high-voltage power supply and metal plate electrodes separated by a thin air gap. A high-frequency alternating current of several thousand volts is applied to the electrodes to establish a corona discharge. A flow of dry air or oxygen from an external source is passed between the electrodes where the electrical discharge disrupts some of the oxygen molecules and generates ozone. To improve efficiency, the electrodes are usually coated with a dielectric material to ensure uniform current flow over the surface of the plate. A large amount of heat is generated by the electric field and the dielectric material must have both high electrical resistance and high thermal conductivity, two properties that do not normally occur together. The usual design compromise is to use a thin layer of a highly resistive dielectric material to compensate for the reduced heat conductivity.

The efficiency of ozone generation can be controlled by the voltage, waveform, and frequency of the electrical power applied to the electrodes and by the gas flow rate through the generator. At a given voltage, higher frequencies (up to about 1,000 Hz) result in increased ozone production. Thus, the frequency is usually fixed at about 1,000 Hz and the ozone production rate is controlled by vary-

29. Primarily NO (with some NO_2 and N_2O_5) formed from the reaction of N_2 and O_2 in the air passing through the corona discharge.

ing either the voltage or the gas flow rate. Because the efficiency of ozone generation drops off rapidly as the gas flow increases, the flow rate is normally fixed at the lowest practical rate and the amount of ozone produced is controlled by varying the voltage. Under ideal conditions, the maximum attainable efficiency is about 25 grams of ozone per watt-hour of electrical power consumed (Rosenthal and Wilson 1987). This translates into ozone concentrations on the order of 10% O_3 by weight in the outlet stream when oxygen is used as the feed gas and 5% when air is used. In practice, 8% ozone from oxygen gas and 3% ozone from air are more realistic efficiency levels. As mentioned, a considerable amount of the electrical power applied to the generator is lost as heat even at the highest efficiency level. Air or even water cooling is always needed, depending on ozonator output. As a guideline, a water-cooled generator with a rated output of 100–200 lbs O_3 per day is usually sufficient to disinfect each 5,000–10,000 gpm of a hatchery water supply.

As mentioned, either oxygen or air can be used as the feed gas. Although air is essentially free and oxygen is somewhat expensive, it is often cheaper to use oxygen, particularly in larger generators, because considerably less electrical power is required to generate the same amount of ozone. Whatever gas is used, it must be dry and free of dust and oil. A moisture content equivalent to a dew point of –40°C is the maximum safe level and –60°C is recommended. Ozone is highly corrosive alone, but mixed with air containing only a few ppm moisture, it will soon destroy the generator. With small ozonators (e.g., those employing UV lamps used in home aquaria), it is less expensive to replace the ozonator periodically than to dry the inlet air.

Ozone is a very effective disinfectant. Like chlorine, its effectiveness is a function of contact time and concentration. However, ozone is a stronger oxidant than chlorine and fish pathogens can generally be inactivated with much shorter contact times—often in a matter of seconds or minutes. In contrast to chlorine, ozone disinfection is much less affected by the water pH (over the fish cultural range of 6–9) and its germicidal efficacy is not reduced by reaction with ammonia to form chloramines. However, unlike chlorine which readily dissolves in water and then reacts with it to form germicidal hypochlorous acid, the solubility of O_3 is limited and physical contact between the target organisms and microbubbles of ozone gas in the water is an important mechanism of action. Thus, the efficiency of gas dispersion and mixing will affect the ozone dose and contact time required to inactivate fish pathogens.

Ozone–water contactors practical for fisheries applications fall into three groups: (1) spray towers, (2) packed columns, and (3) ozone gas dispersion directly into the water through airstones. Ozone–water contact in its simplest form is bubbling ozone gas through a micropore diffuser into a column of water. Unlike the oxygen aeration systems previously discussed, carbon rod diffusers cannot be used because they would quickly be oxidized. However, suitable ceramic and sintered glass spargers are commercially available in a wide variety of pore sizes and physical shapes. Venturi aspirators can also be used for mixing ozone and water. For fisheries/aquaculture applications however, ozone gas dispersion through ceramic diffusers is usually adequate. Details of spray tower and packed column contactor design can be found in Stahl (1975).

To use ozone most efficiently, it must be recognized that both organic and inorganic matter in hatchery water supplies will exhibit an ozone demand and increase the O_3 concentration needed for disinfection. Ideally, settling or filtration should be used to reduce the turbidity due to suspended sediments to 20 NTU or less. If this is not practical, a compensatory increase in ozone concentration can be used to achieve the required degree of disinfection. In contrast to breakpoint chlorination, however, stable ozone residuals may be difficult to achieve in some natural waters even though microorganisms are rapidly being killed. In these cases, a biological end point, such as a 5-log reduction in pathogen numbers, should be used instead of the more conventional chemical measurement. For convenience, the biological end point chosen can be referenced to a chemical measurement such as the ozone application rate needed to achieve an equivalent residual in ozone demand-free water (Wedemeyer and Nelson 1977).

Inorganic cations such as dissolved iron and manganese also exhibit an ozone demand. Surface waters used for (salmonid) fish culture are usually low in such dissolved minerals but high concentrations of undesirable cations often occur in groundwater supplies. Iron, in particular, must often be removed from well water before it can be used for aquaculture. Oxygen aeration to oxidize dissolved cations to insoluble hydroxides that can be precipitated by chemical flocculation is the conventional method. However, ozonation will produce insoluble metal oxides and can be a more effective removal method. In one such case, ozonation reduced the dissolved iron concentration from its initial concentration of 9.5 mg/L down to 0.07 mg/L, and the manganese concentration from 1.2 mg/L to 0.05 mg/L (Wheaton 1977). Conventional oxygen aeration only reduced the iron concentration to about 4 mg/L and the manganese to 0.7 mg/L. In contrast, ozone oxidation of inorganic cations can be a disadvantage, especially in closed marine shellfish culture systems, because dissolved salts required for metabolism may also be oxidized to insoluble forms, making them biologically unavailable.

Although ozone can be effective for such purposes as iron oxidation, the depuration of coliform contaminated shellfish, and the inactivation of paralytic shellfish poisoning (PSP) toxins, its primary fisheries application continues to be fish disease control. As with the previously discussed disinfectants, the specific ozone concentrations and contact times required to inactivate fish pathogens varies with the target organism and water quality. In brackish or marine waters, for example, ozonation cogenerates germicidal concentrations of bromine gas (Br_2) and hypobromous acid (HOBr) from the high levels of dissolved Br^- present. In seawater aquaculture systems, Blogoslawski and Stewart (1977) reported effective bacterial kills at an ozone concentration of about 0.5 mg/L. Activated carbon filters were used to remove residual ozone oxidants such as Br_2. Based on kill times for common fungi, bacteria, virus, and protozoan parasite species, an ozone dosage between 0.1 and 0.5 mg/L with 5–10 minutes of contact time will provide a conservative margin of safety when disinfecting freshwater hatchery supplies with low suspended solids levels (<20 NTU) (Owsley 1991). Maintaining an O_3 concentration of about 0.03 mg/L provides effective *Saprolegnia* control during (salmonid) egg incubation (Wedemeyer et al. 1979). These recommendations should be treated as guidelines only. Dose response relationships should always be developed for the specific water to be treated. Because ozone disinfection is not limited to low-turbidity water, a results-oriented end point (e.g., a 5-log reduction in pathogen numbers) may be needed for hatchery wastewater treatment because a measurable residual may be difficult to detect if high sediment levels are present. In such cases, the instantaneous O_3 concentration may be estimated or the O_3 application rate may be referenced instead.

If spores are present, the foregoing guidelines must be further modified. For example, spores of *Bacillus polymyxa* can resist ozone concentrations of 1 mg/L for 10 min in freshwater while zoospores of the abalone parasite *Labyrinthuloides haliotidis* can survive exposure to 0.3 mg/L O_3 for 19 min in seawater (Colberg and Lingg 1978; Bower et al. 1989). Thus, on-site tests should always be conducted to confirm that the desired reduction in pathogen load is actually being achieved. A summary of ozone dose and contact times recommended for general fishery applications is given in Table 6.3.

Methods for Removing Ozone. As with chlorination, the residual ozone remaining in the water after the target microorganisms have been inactivated is highly toxic to fish and must be removed down to a concentration of <0.002 mg/L before the water enters a rearing unit (Wedemeyer et al. 1979). Fortunately, ozone is naturally unstable in water[30] and decomposes with a half-life of 10–20 minutes into molecular oxygen. Aeration or mechanical aeration can shorten this process to a few minutes and, in smaller scale systems, may be sufficient to prevent fish toxicity. In hatcheries where several thousand gpm of water must be deozonated, packed column stripping towers similar to those used for degassing supersaturated water are very effective. A 3-m length of 76-cm diameter pipe packed with 6-cm biofilter (Koch) rings will remove 95% or more of the residual ozone from a water flow of about 20,000 Lpm if a countercurrent flow of forced air is provided (Owsley 1991). Further treatment to remove ozone down to the 0.002 mg/L level chronic safe exposure level (for salmonids) may not be required. However, if the stripping towers are installed inside hatchery buildings, ventilation to remove the off-gas will probably be required to keep ozone in the air below 0.1 mg/L—the

30. Ozone in air is much more stable and traces of O_3 may persist for hours instead of minutes.

TABLE 6.3. Recommended ozone dose and contact times to inactivate specific fish pathogens and for general aquaculture applications.

Pathogen/Application	Dose (mg/L)	Contact Time	Comments
	Pathogen Inactivation		
Myxobolus cerebralis	0.3	10 min	Turbidity <10 NTU
Ceratomyxa shasta	1–2	30 min	
Saprolegnia sp.	0.03	continuous	salmonid egg incubation
Aeromonas salmonicida	0.1	1 min	
IHNV, IPNV	0.01	1 min	Turbidity <10 NTU
	Fishery Applications		
Hatchery water supply (5,000–10,000 gpm)	0.1–0.5 mg/L	5–10 min	Turbidity <20 NTU
Iron removal from well water supplies	0.5 mg/L	10 min	

Source: Compiled from Wedemeyer et al. (1979); Rosenthall and Wilson (1987); Tipping (1988).
Note: Dose-response relationships for the particular water to be treated should always be developed. Ozone must be removed from the water before it contacts fish. The maximumun safe chronic exposure level (for slamonids) is 0.002 mg/L.

safe upper limit for human respiration. Alternatively, the off-gas can be passed through commercially available catalytic heaters to decompose the traces of ozone that remain.

Activated carbon filtration and sodium thiosulfate additions can also be used to remove residual ozone oxidants from treated freshwater and seawater (Owsley 1991). Carbon filtered ozonated seawater is nontoxic to embryos and larvae of the American oyster (*Crassostrea virginica*) (Blogoslawski and Stewart 1977). Sodium thiosulfate should be added at 1 mg/L to neutralize residual concentrations of 0.2 mg/L ozone oxidants in seawater aquaculture systems (Hemdal 1992). With either of these methods, there is no off-gas to deal with.

REFERENCES

Batts, W. N., M. L. Landolt, and J. R. Winton. 1991. Inactivation of infectious hematopoietic virus by low levels of iodine. Applied and Environmental Microbiology 57:1379–1385.

Blogoslawski, W. J., and M. E. Stewart. 1977. Marine applications of ozone water treatment. Pages 226–276 in C. Fochtman, R. Rice, and M. Browning (eds.), Forum on Ozone Disinfection. International Ozone Institute, Syracuse, New York.

Bower, S. M., D. J. Whitaker, and D. Voltolina. 1989. Resistance to ozone of zoospores of the thraustochytrid abalone parasite, *Labryrinthuloides haliotidis*, (Protozoa: *Labrithromorpha*). Aquaculture 78:147–152.

Caufield, J. D. 1991. Specifying and monitoring ultraviolet systems for effective disinfection of water. American Fisheries Society Symposium 10:421–426.

Chick, H. 1908. An investigation into the laws of disinfection. Journal of Hygiene 8:92–158.

Colberg, P. J., and A. J. Lingg. 1978. Effect of ozonation on microbial fish pathogens, ammonia, nitrate, nitrite, and BOD in simulated reuse hatchery water. Journal of the Fisheries Research Board of Canada 35:1290–1296.

Dey, D. B., and D. M. Damkaer. 1991. Effects of spectral irradiance on the early development of chinook salmon. Progressive Fish-Culturist 52:141–154.

Hemdal, J. F. 1992. Reduction of ozone oxidants in synthetic seawater by use of sodium thiosulfate. Progressive Fish-Culturist 54:54–56.

Hoffman, G. L. 1975. Whirling disease (*Myxosoma cerebralis*): control with ultraviolet irradiation and effect on fish. Journal of Wildlife Diseases 11:505–507.

Jeffrey, D. J. 1995. Chemicals used as disinfectants: active ingredients and enhancing additives. Scientific and Technical Reviews, Office International des Epizooties 14:57–74.

Jensen, G. 1989. Handbook for common calculations in finfish aquaculture. Publication 8903 Louisiana State University, Agricultural Center. Baton Rouge, Louisiana.

Kerr, J. B., and C. T. McElroy. 1993. Evidence for large upward trends of ultraviolet-B radiation linked to ozone depletion. Science 262:1032–1034.

Owsley, D. E. 1991. Ozone for disinfecting hatchery rearing water. American Fisheries Society Symposium 10:417–420.

Pascho, R. P., M. L. Landolt, and J. E. Ongerth. 1995. Inactivation of *Renibacterium salmoninarum* by free chlorine. Aquaculture 131:165–175.

Rosenthal, H., and J. S. Wilson. 1987. An updated bibliography (1845–1986) on ozone, its biological effects and technical applications. Canadian Technical Report of Fisheries and Aquatic Sciences No. 1542. Department of Fisheries and Oceans, Halifax, Nova Scotia, Canada.

Seegert, G. L., and A. S. Brooks. 1978. Dechlorination of water for fish culture: Comparison of the activated carbon, sulfite reduction, and photochemical methods. Journal of the Fisheries Research Board of Canada 35:88–92.

Sharp, D. G., and J. Leong. 1980. Inactivation of poliovirus I (Brunhilde) single particles by chorine in water. Applied Environmental Microbiology 40:381–385.

Stahl, D. E., 1975. Ozone contacting systems. Pages 40–55 *in* First International Symposium on Ozone for Water and Wastewater Treatment. International Ozone Institute, Waterburry, Connecticut.

Tipping, J. 1988. Ozone control of ceratomyxosis: survival and growth benefits to steelhead and cutthroat trout. Progressive Fish-Culturist 50:202–210.

Torgersen, Y., and T. Håstein 1995. Disinfection in aquaculture. Scientific and Technical Reviews, Office International des Epizooties 14:419–434.

Ward, H. M. 1893. The action of light on bacteria. Proceedings of the Royal Society of London 54:472–489.

Watson, H. E. 1908. A note on the variation of the rate of disinfection with change in the concentration of the disinfectant. Journal of Hygiene 8:536–542.

Wedemeyer, G. A., and N. C. Nelson. 1977. Survival of two bacterial fish pathogens (*Aeromonas salmonicida* and the Enteric Redmouth Bacterium) in ozonated, chlorinated, and untreated waters. Journal of the Fisheries Research Board of Canada 34:429–432.

Wedemeyer, G. A., N. C. Nelson, and W. T. Yasutake. 1979. Potentials and limits for the use of ozone as a fish disease control agent. Ozone: Science and Engineering 1:295–318.

Wei, J. H., and S. L. Chang. 1975. A multi-Poisson distribution model for treating disinfection data. Pages 11–47 *in* J. Johnson (ed.), Disinfection of Water and Wastewater. Ann Arbor Science Publishing, Ann Arbor, Michigan.

Wheaton, F. W. 1977. Aquacultural Engineering. John Wiley, New York.

Yoshimizu, M., H. Takizawa, M. Sami, H. Kataoka, T. Kugo, and T. Kimura. 1990. Disinfectant effects of ultraviolet irradiation on fish pathogens in hatchery water supply. Pages 643–646 *in* R. Hirano and I. Hanyu (eds.), Second Asian Fisheries Forum. Asian Fisheries Society, Manila, Philippines.

Index